Closing the Robotics Market Gap through the Robotics for Inspection and Maintenance Network

Closing the Robotics Market Gap through the Robotics for Inspection and Maintenance Network

Edited by

Christophe Leroux

Cambridge Scholars Publishing

Closing the Robotics Market Gap through the Robotics for Inspection and Maintenance Network

Edited by Christophe Leroux

This book first published 2023

Cambridge Scholars Publishing

Lady Stephenson Library, Newcastle upon Tyne, NE6 2PA, UK

British Library Cataloguing in Publication Data
A catalogue record for this book is available from the British Library

ISBN (10): 1-5275-4665-9
ISBN (13): 978-1-5275-4665-3

TABLE OF CONTENTS

LIST OF FIGURES

LIST OF TABLES

SECTION A

THE CURRENT STATE OF AFFAIRS

CHAPTER 1

INTRODUCTION

CHRISTOPHE LEROUX

1.1. EU infrastructure

> *"Infrastructure consists of the basic assets and structures that support economic activity" to quote a report from EPRS[1] a think tank of the European Parliament. The report further adds that, "Investment in such assets is* ***markedly different from other types of capital expenditure****, due to the* ***heavy involvement of the public sector*** *and the* ***significant positive spill-over*** *that it generates throughout the economy".*

Investment in infrastructure in the key sectors of energy production and distribution, transport, water supply and sanitation, and telecommunications are fundamental to European society, which needs new technologies and modernized infrastructures to cope with population growth and demographic changes. The European Investment Bank (EIB) estimates indicate that achieving the targets of the EU digital agenda would demand additional investment of approximately €55 billion per annum.

According to the EPRS study[1], investment in infrastructure has been declining since 2009 in Europe. EIB estimated that investment needs for energy, transport, water supply and sanitation, and telecoms are as much as €688 billion per year. For water and sanitation, and telecoms, the needs reach €160 billion and €138 billion respectively. In addition, EIB estimates current investment in social infrastructure (health, education and social housing) at €142 billion per year[2] in the EU.

The investment needed for the inspection, maintenance and repair of infrastructure activities is equiparable to those required for the construction of the infrastructure at any given time. Experiences with roads show that if maintenance is neglected over a period of 3 years, the necessary repairs to these roads will probably cost 3 to 6 times more than relevant maintenance[3]. A Chicago Metropolitan Planning Council report[4] indicated that maintenance accounts for the bulk of its resources, nearly 90% of the more than $2 billion the state spends each year on highways. Poor infrastructure maintenance can also reveal or cause fears and changes in behaviour – for example, a J.D. Power survey[5] shows that one quarter of Americans claim never to drink tap water due to the bad smell and fears about water safety. In the USA, local and state governments collectively spend in excess of $100 billion per year on the construction, maintenance, and repair of roads, bridges, and ancillary components (e.g., FHWA 2008, 2019)[6]. Different analyses of the importance of inspection and

---

1 Zachariadis Ioannis Agamemnon, Investment in infrastructure in the EU Gaps, challenges, and opportunities, October 2018. EPRS | European Parliamentary Research Service
https://www.europarl.europa.eu/RegData/etudes/BRIE/2018/628245/EPRS_BRI(2018)628245_EN.pdf
2 EIB, Investment Report 2017/2018: From Recovery to Sustainable Growth, November 2017.
3 Burningham, Sally; Stankevich, Natalya. 2005. « *Why Road Maintenance is Important and How to Get it Done* ». Transport Notes Series; No. TRN 4. World Bank, Washington, DC. © World Bank.
https://openknowledge.worldbank.org/handle/10986/11779
4 Peter Skosey Chicago Metropolitan Planning Council, "*Fix it first: The importance of infrastructure maintenance* », 2012, https://www.metroplanning.org/news/6578/Fix-it-first-The-importance-of-infrastructure-maintenance
5 J.D. Power Finds, *Results of poor infrastructure maintenance reveals fears and changes in behaviour*, May 2020, https://www.jdpower.com/business/press-releases/2020-us-water-utility-residential-customer-satisfaction-study
6 Mc Donough Ryan P., Yan Claire J., *Maintaining Maintenance: The Real Effects of Financial Reporting for Infrastructure,* July 2021,
https://www.brookings.edu/wp-content/uploads/2021/05/McDonough-Yan-Paper_July2021.pdf

maintenance can be found in various reports[7]. Figure 1 illustrates the links between the cost of maintenance and the amount of maintenance.

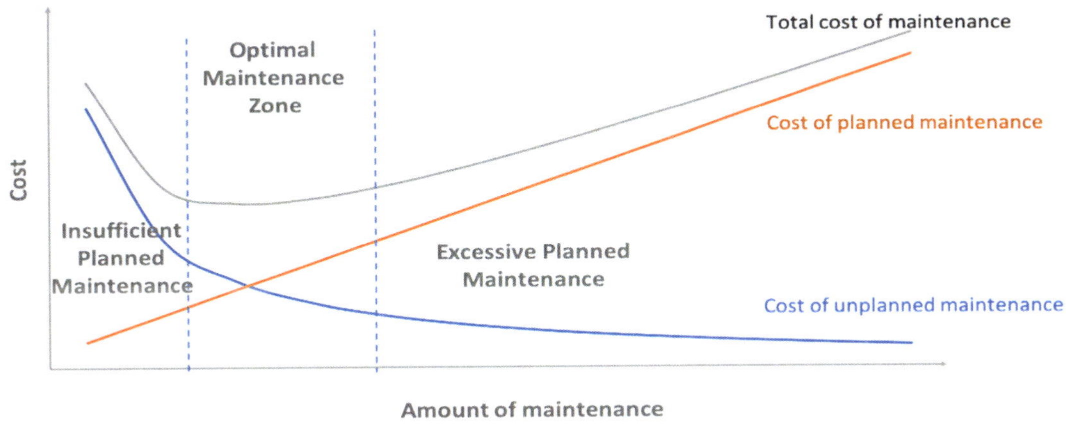

Figure 1: Amount of maintenance[8]

This book presents the efforts of the RIMA (Robotics for Inspection and Maintenance) European H2020 project9 to stimulate innovation in the use of robotics technologies for the inspection, maintenance and repair of infrastructure. The aim of RIMA was to create an instrument—a network of Digital Innovation Hubs (DIHs) on inspection and maintenance (I&M) robotics named the RIMA Network—and to demonstrate the usefulness of this network in funding and mentoring small and medium-sized companies (SMEs) that are developing new solutions for the inspection and maintenance of infrastructure in different sectors.

We present a summary of the study we conducted to analyse the market potential of robot technologies that inspect and maintain infrastructure. The results of this study were used to frame the project. We present the concept of DIHs and the way that RIMA activities and services were organized based on the principle of supporting SMEs willing to create innovative solutions. We also present details of the competitive cascade funding (FSTP) instrument that was used to support SMEs financially, technically and in the development of their product and their business. The SMEs that won this funding ran experiments mentored by RIMA partners who also provided services. We introduce the RIMA Alliance that was created out of the RIMA Network in order to promote the sustainability of our network and engage further organizations interested in supporting innovation in robotics for I&M. This book finishes by providing our recommendations based on lessons learned throughout our RIMA experience.

REFERENCES FOR FURTHER INFORMATION

Abiad A., Almansour A., Furceri D., Mulas Granados C., and Topalova P., *Is it time for an infrastructure push? The macroeconomic effects of public investment*, IMF, October 2014. EPRS | European Parliamentary Research Service

Burningham, Sally; Stankevich, Natalya. 2005. « Why Road Maintenance is Important and How to Get it Done ». Transport Notes Series; No. TRN 4. World Bank, Washington, DC. © World Bank. https://openknowledge.worldbank.org/handle/10986/11779 License: CC BY 3.0 IGO

7 Babych Yasya, Leruth Luc, *For a Better Budget Management of Infrastructure Investments*, Freenetwork, February 2021, https://freepolicybriefs.org/wp-content/uploads/2021/04/freepolicybriefs20210419.pdf

8 Thi Hoai Le, A., N. Domingo, E. Rasheed, and K. Park, "Building Maintenance Cost Planning and Estimating: A Literature Review," 34th Annual ARCOM Conference, Belfast, UK (2019).

9 Project funding page: https://cordis.europa.eu/project/id/824990

Council of Europe Development Bank, *Investing in public infrastructure in Europe: A local economy perspective*, February 2017.

European Commission, *Infrastructure in the EU: Developments and impact on growth, European Economy*, Occasional Papers 203, December 2014. European Energy Security Strategy, COM (2014)330, European Commission, May 2014.

European Commission, *White paper on transport. Roadmap to a Single European Transport Area - Towards a competitive and resource efficient transport system*, March 2011.

EIB, Investment Report 2017/2018: *From Recovery to Sustainable Growth*, November 2017.

EIB, Restoring EU competitiveness, January 2016.

Mc Donough Ryan P., Yan Claire J., *Maintaining Maintenance: The Real Effects of Financial Reporting for Infrastructure,* July 2021, https://www.brookings.edu/wp-content/uploads/2021/05/McDonough-Yan-Paper_July2021.pdf

Szczepański, M., Public investment to support long term economic growth in the EU, EPRS, European Parliament, July 2016.

1.2. The field of inspection and maintenance

Inspection and maintenance (I&M) represents a large economic activity spanning across multiple sectors such as energy production and distribution, water supply, transport, and communication. There is massive potential for robotic applications, to increase productivity, reduce costs and improve safety. The total market size of robotics is however still negligible in relation to the overall market size.

Currently the EU is a global leader in this rapidly growing field with more than half of all manufacturers of I&M robots based in Europe. The EU hosts over 50% of I&M robotics offer but there is a bottleneck connecting these technologies to the market and applications with high potential.

RIMA was a 4-year project that aimed to tackle this bottleneck by establishing a network of DIHs and industry / industrial associations (the "RIMA Network") to support the uptake of robotics – and help SMEs to develop novel I&M solutions for different sectors. The support from sectorial associations was essential for making a bridge with relevant industrial actors in the value chain.

Challenges included: reinforcing network connections, providing education and training about I&M using robotics, and connecting the value chain—researchers, technology providers, service providers, asset owners, asset operators, and investors – in order to accelerate economic growth in the field.

RIMA gathered leading research organizations from across Europe to create a network, each research organization supports a regional DIH which has activities aligned with regional policies and industrial sectors. The RIMA Network was built in close cooperation with the SPRINT robotics association, a project partner, in order to ensure the sustainability of the network.

RIMA support to SMEs is organized using a bottom-up approach driven by end-user needs and presented in the form of a set of services designed to facilitate understanding and dialogue across all relevant actors in the value chain.

More than 2/3 of RIMA's budget was used to support SMEs financially and mentor SME experiments

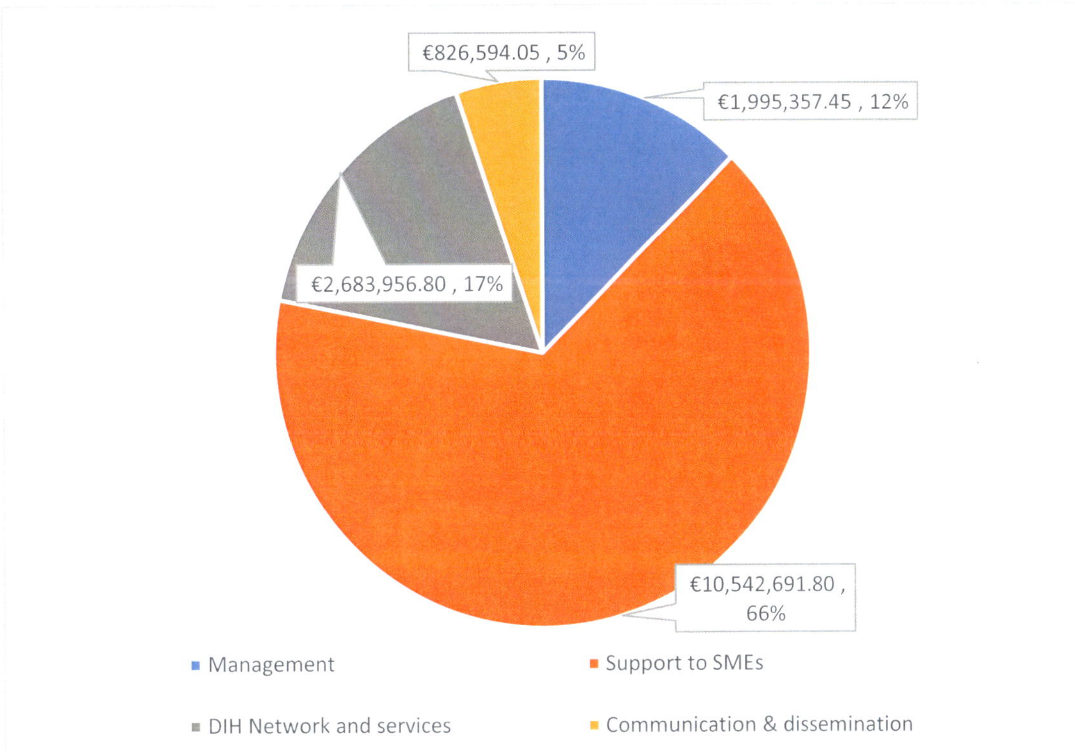

Figure 2: Budget share per activity in RIMA

The objectives of RIMA Network were

- increased competitiveness of EU I&M robotics,
- economic added value by increased productivity and availability of the critical infrastructure,
- social and environmental impact through improved safety and reduced emission of hazardous substances, and
- the constitution of a sustainable and scalable (open to new members) DIH network aligned with industrial and European policies and ambitions.

SECTORS COVERED

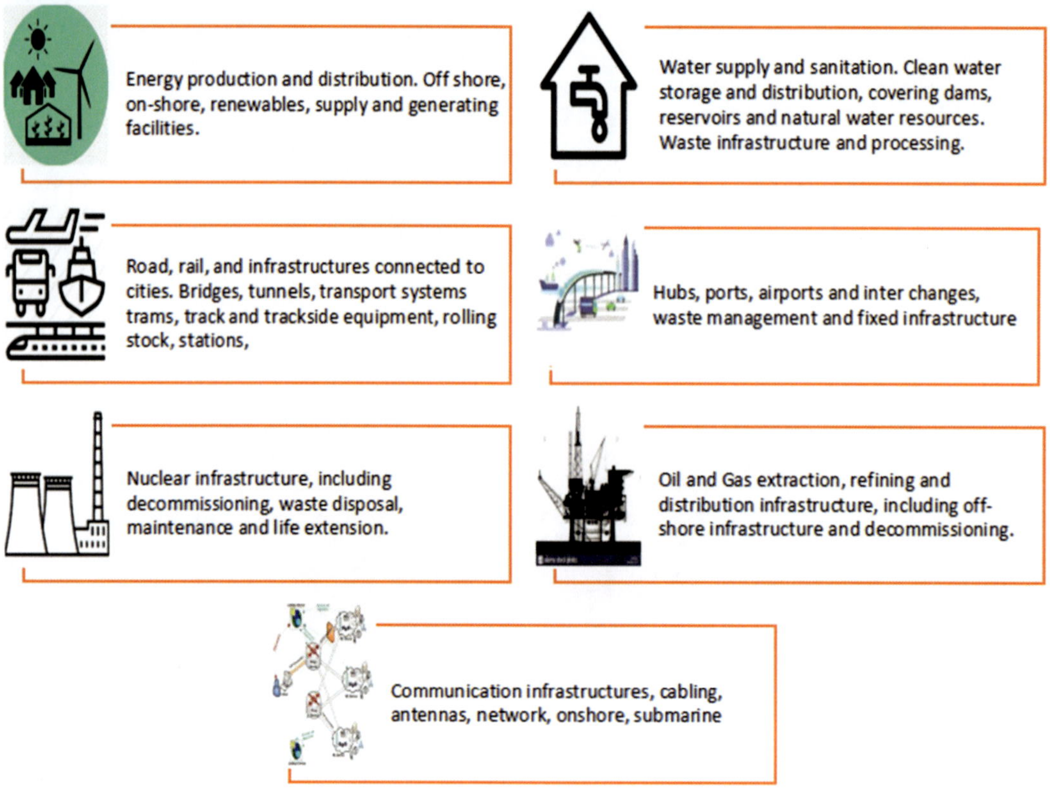

Figure 3: Sectors covered by RIMA Network members

CHAPTER 2

THE RIMA NETWORK

BEGOÑA ARRUE CHIQUINQUIRA AND JOSÉ LUIS ANDRADE PINEDA

2.1. RIMA Network core features

RIMA COMMUNITY TOOL

RIMA Network seeks to connect and inspire key stakeholders in I&M robotics and to accelerate innovation and uptake of robotics. The RIMA Network Community is a place to collaborate, share information about I&M robotics and keep up-to-date about RIMA activities.

In this way, the RIMA Network aspires to bring together technology providers, service providers, asset owners and operators, DIHs, and other facilitators, to join forces and competences in accelerating robotics deployment in I&M.

Join us to be part of this global community to collaborate, share knowledge, and participate in RIMA events and activities!

FUNDING

RIMA provided financial and technical support for 50 cross-border experiments involving European SMEs through 2 Open Calls (2019 / 2020) with a total funding of 8.1M€.

Successful candidates received equity-free funding for technology transfer and development in I&M robot applications.

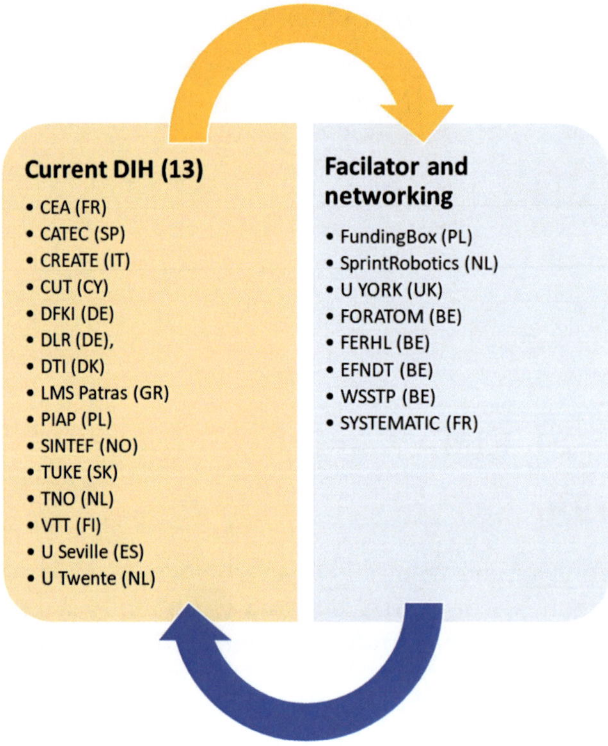

Figure 4: The RIMA Network at project start

NETWORK

RIMA brought DIHs and other facilitators together into a common network that allows them to join forces and competences in promoting I&M robotics. The facilitator organisations have the connections and network to reach a broad set of industries in their domains of expertise. The initial RIMA Network comprised 13 DIHs and 8 facilitator organisations and is still open to new members through the RIMA Alliance framework.

DIHs offer key acceleration services for companies. These DIHs include organizations involved in research and innovation activities, and they offer expertise in robot technologies, sensors, and other equipment and services applicable for I&M operations.

TRAINING

The RIMA Network provides skills and tools to implement the latest methods in knowledge transfer between academia and industry.

The RIMA training framework aims to increase digital skills and knowledge of robotics, especially within the I&M domain, amongst a wide range of people – from students starting out on their career path, through the current workforce who need to retrain, managers who need to understand the importance and effect of digitisation, and start-up technology developers that need concrete information and support for getting their upcoming technology to market.

Examples of services

- •Advising: finances, technologies,
- •Proof of concept
- •Support to tech transfer
- •Support for testing
- •Access to Incubators
- •Access to actors of value chain
- •Training, coaching,
- •Path to market, market deployment
- •Communication, dissemination

Figure 5: Examples of RIMA Network services

2.2. The concept of DIHs

Digital Innovation Hubs are one-stop-shops that help companies become more competitive with regard to their business/production processes, products or services using robotics and other digital technologies. DIHs provide access to technical expertise and experimental facilities, so that companies can "test before invest". They also provide innovation services—such as financial advice, training and skills development—that are needed for a successful digital transformation. Key environmental issues are also considered, in particular regarding energy consumption and low carbon emissions. As proximity is considered crucial, they act as a regional first point of contact, a doorway for local organisations to reach through, and thereby strengthen the local innovation ecosystem. A DIH is a regional cooperation between organisations such as research and technology organisations (RTOs), universities, industrial associations, incubators/accelerators, chambers of commerce, regional development agencies and vocational training institutes and can also have strong connections with service providers located outside their region, supporting local companies by providing access to services and facilities that are not available locally.

Each DIH acts as the epicentre of a local/regional or even national digital innovation ecosystem able to provide access to services, facilities, and expertise belonging to a wide range of organisations. The aim is to ensure that the individual customers (mostly from smaller companies or the public sector) get the services they need and that regional market segments get access to innovative, scalable solutions.

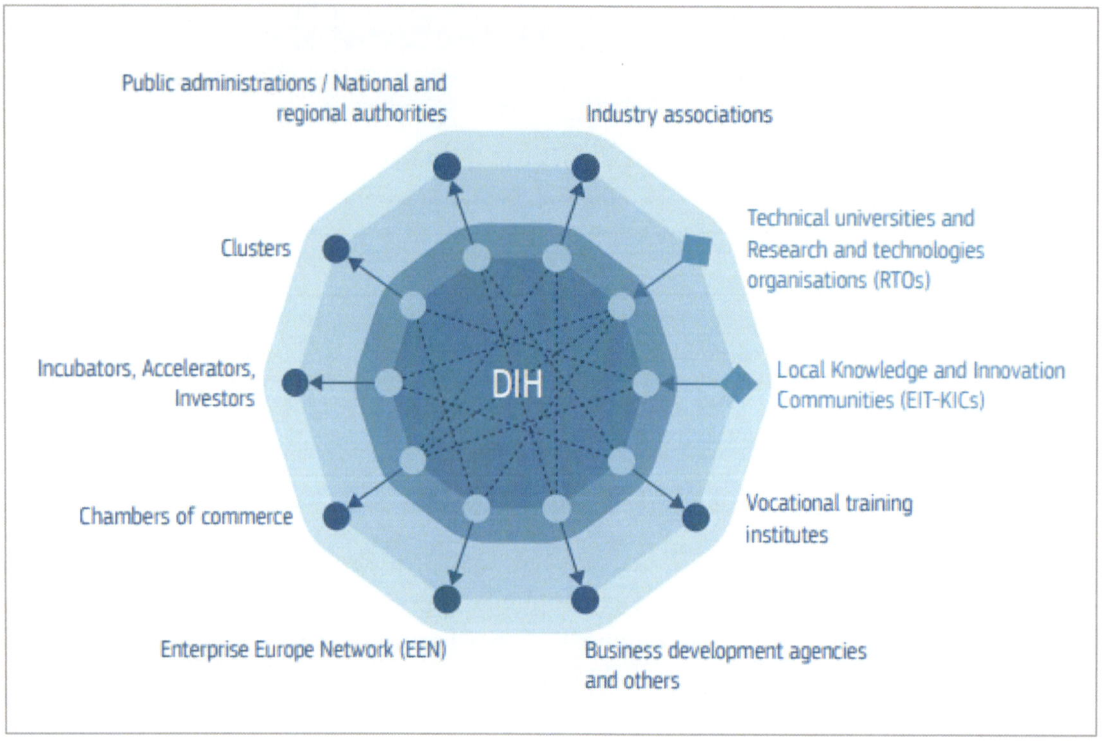

Figure 6: DIH concept

To reach that goal, DIHs cooperate with each other at regional, national and/or European level. Customers can approach different DIHs depending on their technological and sectorial specifications. Any organisation interested in DIH services can choose the DIH that best matches their criteria (geography, sector, technology, type of service, etc.) via an interactive tool that helps navigate the network of DIHs available in Europe.

The RIMA Network maximises synergies among its members and contributes to promoting a joint approach and branding, making possible a bigger recognition of the initiative outside its ecosystem. Technology supplier and developer involvement means that RIMA has access to a wide range of expertise and facilities. Our ability to reach end users increases the chances of engaging SMEs and other potential customers. RIMA activities can also be seen as a way to raise awareness about Europe's "Digitising European Industry" strategy and the role of the DIHs within it. In addition, RIMA Network members have been able to achieve the realisation of research and innovation objectives linked to the transformation of robotics deployment in infrastructure companies and the public sector by addressing these user's specific and general challenges.

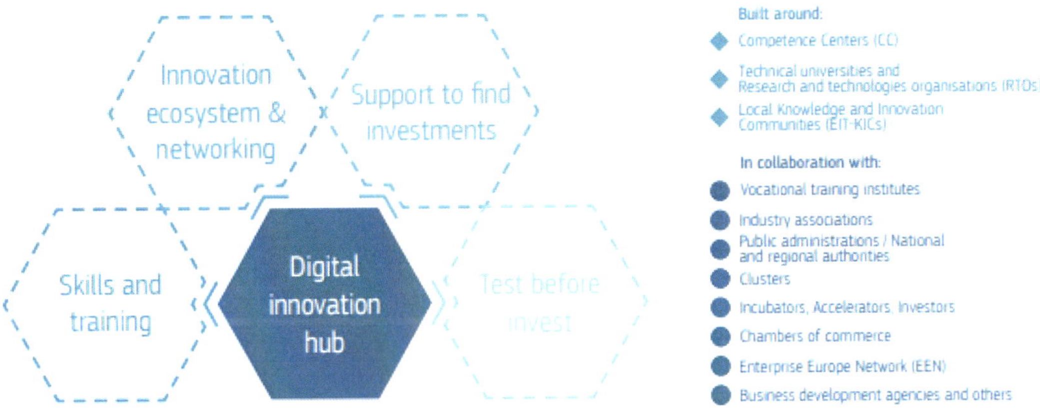

Figure 7: DIHs key information

The main services offered by the RIMA DIHs are:

Access to innovation and leading-edge technologies and applications expertise: in manipulation, teleoperation, navigation, human robot collaboration, supervision, control, mechatronics, sensing, perception, non-destructive testing, AI; for standard and non-standard robots e.g. snake robots, autonomous and semi-autonomous drones: aerial, terrestrial, subsea or underwater, and more.

Funding opportunities and technology acceleration programmes: the Open Calls of RIMA Network Innovation Actions funded experiments and "test before invest" activities to understand the new opportunities available and return on investments, also demonstration facilities and piloting.

Strengthening the relationships of SMEs with DIHs and other relevant actors: through the online community, engagement activities and RIMA training activities.

Skills and training to make the most of digital innovations: "train-the-trainer" programmes, bootcamps, traineeships, exchange of curricula and training material, the RIMA training catalogue.

Support in finding investment and other close-to-market opportunities: supported access to financial institutions and investors in order to access follow-up finance to bring the results of testing and experimentation to the next phase, also access to incubation and acceleration programmes.

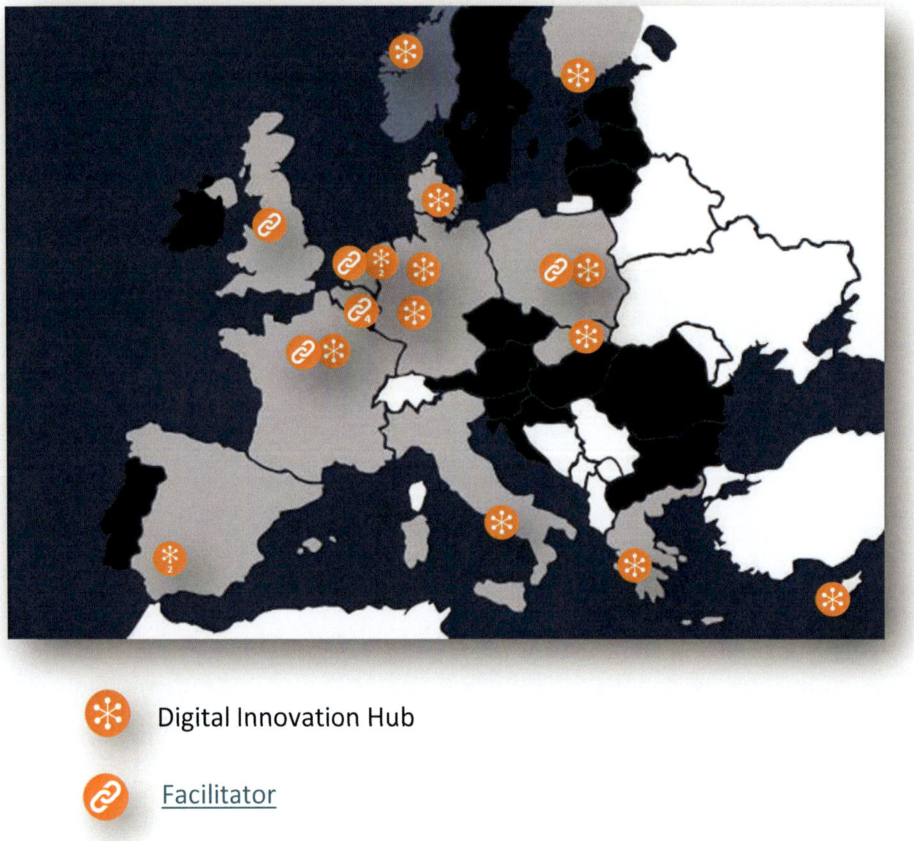

Figure 8: Map of RIMA Network members 2019-2023

To find a full list of European DIHs, not just those in RIMA, see here: https://european-digital-innovation-hubs.ec.europa.eu/edih-catalogue

For a fuller description of the RIMA Network DIHs, please go to section 6.

CHAPTER 3

MARKET BACKGROUND

KRZYSZTOF LIPIEC AND JAKUB GŁÓWKA

3.1. Estimate of the market

The market analysis conducted in early 2021 aimed to frame the innovation activities for Inspection and Maintenance of Infrastructure robotics. The Market Study Unit from CEA conducted this analysis with the support of other RIMA partners.

This section provides key elements contextualizing the positioning of the RIMA DIHs. These elements are available on the community portal. The complete market study is available for free through the SPRINT robotics association[1] as an offer from the network.

In the following subsections, we start with a description of the approach chosen; we then give some figures and some general characteristics of the market. We conclude by presenting the implications of this market research for the value proposition of RIMA's DIHs. This value proposition is transferable to other DIH networks.

THE APPROACH

We had two options for this market study: top down or bottom up. These two options are presented in Figure 9 with some comments about the advantages of each of them.

---

1 https://community.rimanetwork.eu/7423/rimad34v61docx

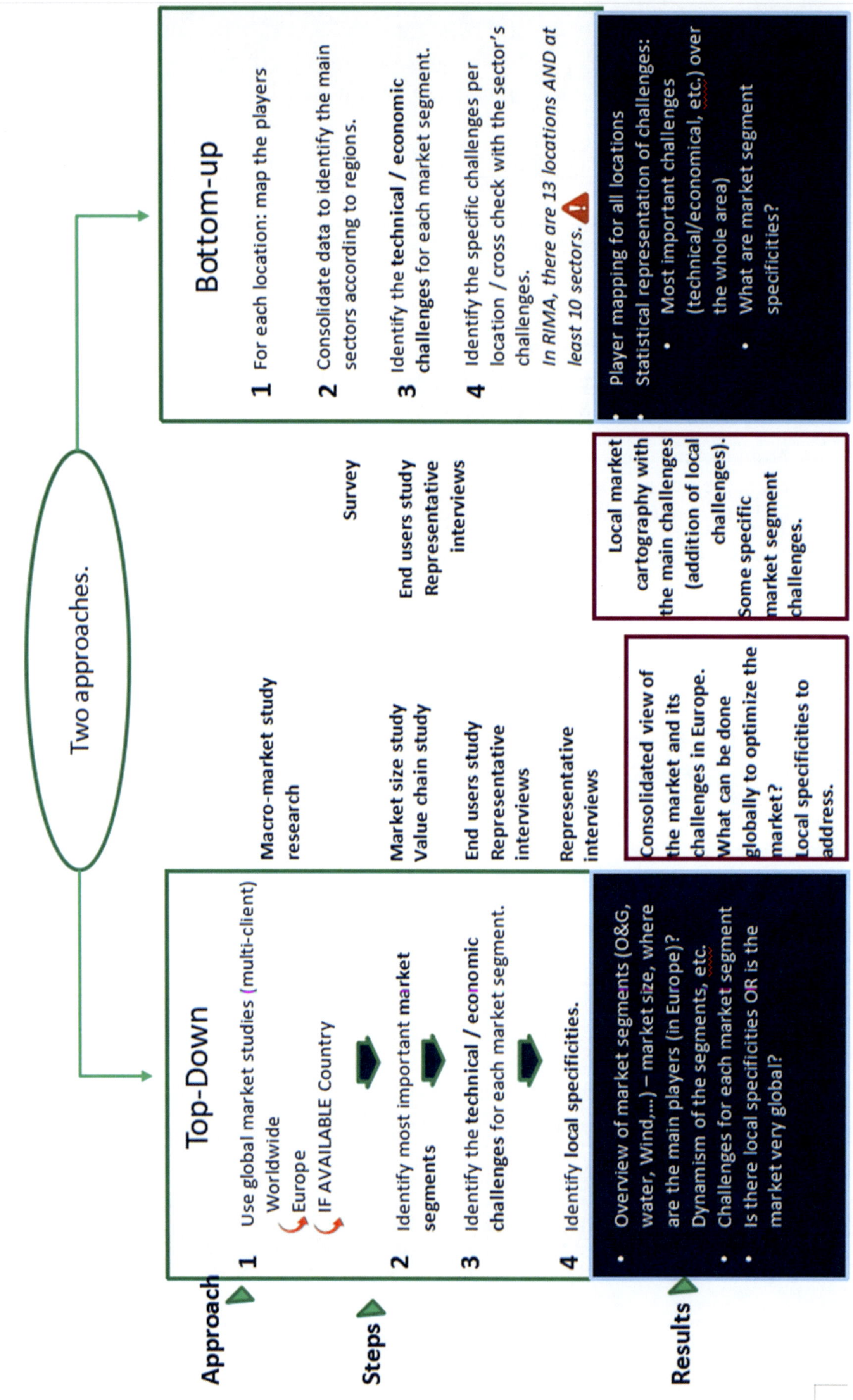

Figure 9: Market study: two possible approaches

The market study started with an investigation of the material already available. No relevant market data was available in the Public Private Partnership SPARC archives, at the EC or via the euRobotics aisbl, and the market information available from the RIMA partners was partial and not uniform. We therefore decided to collect new and unified data. We chose to follow a global approach for the market study for the following reasons.

Added value for RIMA DIHs

Regional DIHs already have good knowledge of their local ecosystem as shown by the activities each DIH has carried out in the past regarding robotics for inspection and maintenance of infrastructure. RIMA partners felt that a global approach would be beneficial for the development of their service provision. The thirteen RIMA DIHs are situated in twelve countries in Europe. Obtaining specific information on markets in regions where no RIMA partner is located would have been very uncertain – there were even difficulties in obtaining reliable information from the regions and countries where RIMA partners were located. Consortium partners also thought that a global market study would help them share a common vision and be more generally relevant.

The multi-sectorial aspect of inspection and maintenance

The RIMA network covers many sectors. Conducting a multi-sectorial market study in as many regions as there are RIMA Network members would have been a considerable effort, and the results would have been difficult to consolidate at European and international scales. Partners' efforts to obtain local market information revealed the challenges of collecting multi-sectorial data and strongly suggested that this approach would not be successful.

Multi-national aspect of some sectors

Some sectors covered by RIMA like oil and gas, nuclear, and renewable energy (wind) are supported by multinationals which are looking for global solutions applicable in all countries over the planet.

Expectations from RIMA customers

All SMEs which interact with the RIMA network require global market data to develop fully.

MARKET ESTIMATES FOR I&M ROBOTICS TECHNOLOGIES

This section attempts to estimate the overal number of potential companies with which RIMA members could interact. These are mostly end-user companies with an interest in using robot technologies to support innovation in infrastructure I&M. The figures below are indicative and will be refined later.

Case of flying drones

The RIMA market study found that around 3,800 companies in France[2] were operating professional drones weighing over 800 g in different sectors in 2019. Europe as a whole is estimated to contain approximately five times the French figure, so extrapolation to Europe gives approximately 3,800*5 = 19,000 companies operating professional drones over 800 g in Europe in different application domains.

The same study indicates that in 2019 the global market for professional services using flying drones for infrastructure I&M involved about 5% of the general market for drones[3]. Making the simplifying assumption that this also corresponds to the number of companies concerned by these subjects, there

2 "Aviation Civile" publication from the DGAC – French Civil aviation authority; 01/2020 DGAC
3 The 2018 Drone Market Sector Report by Skylogic Research LLC

would therefore be 950 companies in Europe operating professional drones in the infrastructure I&M market.

Robots other than drones

The market study also indicates that the I&M market for robots other than drones is at least three times larger than that for drones. Let us assume that the number of companies is proportional to the market, there would therefore be approximately 950*3 = 2,850 companies in Europe operating robots other than drones in the infrastructure I&M market[4].

An overall estimate

Therefore there would be about 2,850 + 950 = 3,800 companies of all sizes operating robots (including drones) for infrastructure I&M.

These estimates are from the first results of the market study. It would not be surprising if in reality there were considerably fewer or, more likely, considerably more companies involved – and the numbers are growing. Reasoning around a range of 1,000 to 10,000 companies offering robotic I&M services (air, sea, submarine, land) would seem be a sound estimate.

Extrapolating the number of "customers" seems riskier. Assuming that each of these companies work for three to five clients maximum, that would mean that there could be approximately 3,000 to 50,000 customers benefiting from the use of I&M robots in Europe.

This estimate should be revisited and consolidated later.

MARKET CHARACTERISTICS

The global inspection and maintenance robotics market in 2020 was worth $4,9bn. Its growth is +19 % CAGR[5].

The I&M robotics market is highly fragmented, both at European and global level. The market is composed of a myriad of companies (around 3,500 as we just mentioned) mostly small and very technical, each offering a very specific robotic platform.

Interviews with stakeholders from different sectors show that robotic companies have to develop a different robot solution for almost every client (the complete market study is provided in a document accessible through the RIMA Network). Figure 10 illustrates a Zion meta study that examined several market investigations[6]. It illustrates market segmentation: the major players for each segment are small and specialist companies that are present in only in one segment. Only a few players appear in several segments.

Although we have seen a few mergers and acquisitions, most robotic companies remain very small, not able to grow because of this lack of market organization. Robotics companies are often incorporated by Non-Destructive Testing (NDT) or inspection services companies (ex: Fugro has completed the acquisition of General Robotics Limited (GRL), Eddyfi Technologies acquired NDT Robotics Leader Inuktun).

---

4 Professional Service Robots Market Size, Share & Trends Analysis Report By Application (Logistics, Healthcare, Customer Service, Field Robots), By Region, And Segment Forecasts, 2020 - 2027 ; Grandview Research 2017
5 https://www.technavio.com/report/inspection-robots-market-industry-analysis
6 Research and markets April 2018, Zion Market Research - April 2018, Transparency Market Research 2018,

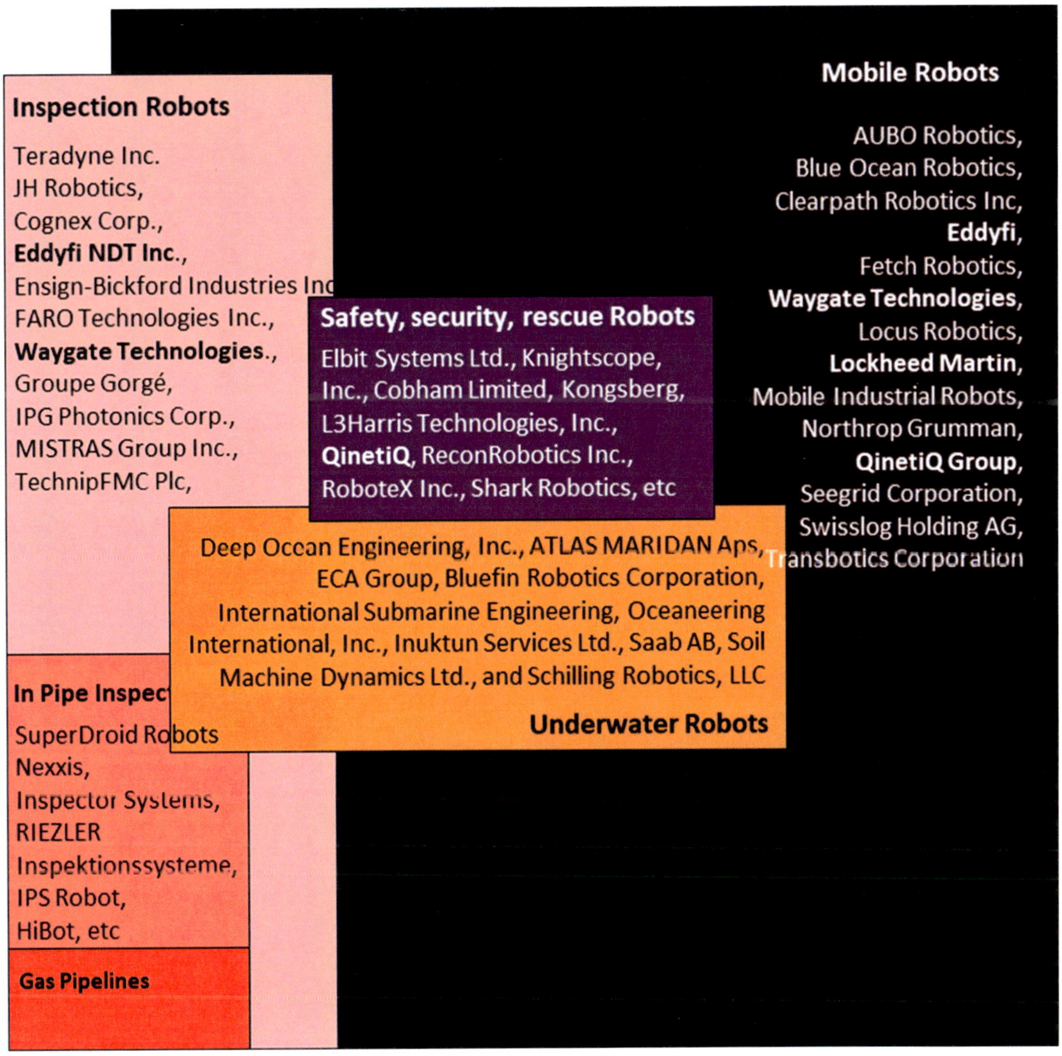

Figure 10: Major players per robotic segment

If we refer to the Technavio market study[13], as a sample a non-exhaustive list of their variety, some of the major players for inspection robots worldwide are depicted in the following picture:

	PERS	REVENUE (2020)	Activity	
EDDYFI	475	130M€	NDT Acquisition of many **pipeline inspection** companies	
TERADYNE (owns Universal Robots)	5 500	$2,1bn	Automatic test equipment	
IPG Photonics Corporation	5 960	$1,4bn (2017)	High-power fiber lasers and amplifiers (inspection, welding, etc)	
Waygate Technologies (a BAKER HUGHES Cie)	300	$38M	Industrial inspection solutions - NDT. Develops mobile inspection robots for **large-scale facilities** in the field of power generation and industrial processing	
JH Robotics	30	$6M	Collaborative robots / automated inspection is a small part of their business	
Cognex Corp.	2000	$725M	Machine vision products / robotic is a small part of their business	
Faro Technologies	1800	$385M	High-precision 3D capture, measurement and analysis across a variety of industries (manufacturing, construction, engineering and public safety)	
Groupe Gorgé	1800	€270M	Safety: Drone and systems, 3D printing and engg	
Nexxis	24	$4M	NDT and Manufacturing automation	

Figure 11: The size of the major companies providing digital I&M technologies, from 13

These example companies are mostly small and most of them are North American. However, field interviews led to an observation that those high-tech companies remain very small. This can be surprising with respect to the high expectations towards robotic and potential of applications. This rises further questions to take into account when approaching these companies, such as: Maybe they are very automated? So employee numbers are expected to be low?

The global market growth for infrastructure I&M robotics, expressed as CAGR, is expected to be around 19%[7]. This is rather good growth with respect to global market growth.

Major market issues

Feedback from robot companies indicates that one of the major issues is the effort needed to develop proof of concepts (PoC) and run tests. Indeed, technology providers have almost to tailor their robots for each of their customers. It is a major trend in this market. Time is dedicated, in particular, to: tuning their robot, finding and/or adjusting testing infrastructures, preparing PoCs, running tests, quantifying and qualifying the results. This work prevents those companies from growing —it is not so much a question of cash, but a matter of resource allocation. Small companies cannot develop several platforms in parallel, therefore they cannot expand their market.

DIH AND RIMA NETWORK VALUE PROPOSITION

The market study conducted during the RIMA project details the challenges that can be addressed by the RIMA network. We focus on two of them.

7 Inspection Robots Market 2020-2024 | Advantages of Robotic Inspection Over Manual Inspection to Boost Growth | Technavio

Need for universality / standardization

As mentioned before, the supply side of the market (SMEs, technology providers) is overwhelmed by a demand that is too scattered. This results in:

- Too much time being spent on the development stage. Preparing PoC is a major activity.
- A poorly consolidated view of the market – the technology providers have only partial knowledge of end-user needs.

The development time can be shortened with the help of RIMA network, for example through assistance for identifying and providing test facilities, through research focused on system or sub-system interoperability, through test standardization, etc.

Consolidation of technical market needs is also a very useful action for a transversal European network. Market investigations or inquiries are examples of actions useful for the stakeholders.

Cross-fertilization between sectors and regions, sharing best practices

Cross-fertilization could be a market accelerator and the role of RIMA would then be to ease the sharing of good practices; and working to overcome reluctance from end users in some sectors to adopt robotic solutions for inspection – some players are however leading the way.

ECOSYSTEM MAPPING

RIMA Network, through its partner, VTT, performed an ecosystem mapping activity in order to represent the complementarities and similarities between various regions as well as highlighting the gaps in the innovation ecosystem. The results were used as the basis for expanding the RIMA network of DIHs, attracting companies to apply for the experiments, and forming the description of the services offered by the RIMA network.

Analysis of the findings of the ecosystem mapping show that majority of the DIHs in the RIMA network are active regionally or nationally. The findings support the assumption that a DIH`s foremost role in the robotics for inspection and maintenance field is being a regional co-operator rather than an international or global actor. However, the DIHs seem to support the regional robotics companies by providing access to service providers and stakeholders within the wider European area. Participation in the ecosystem of multiple DIHs was identified as accelerating this orientation further.

Ecosystem mapping analysis confirms that SMEs are the predominant customer group for most of the DIHs in the RIMA network. A DIH service categorization was created as a result of the RIMA project, including the main categories of 1) ecosystem services, 2) business services, and 3) technology services. We observed that there are not any significant differences between the main categories of services provided by the DIHs in different geographical regions. Stakeholder needs and values concerning robotics for I&M seem to be fundamentally the same at European level. Most DIHs offer a combination of different types of robotics I&M services instead of focussing only on certain service categories. The technology services offered by the RIMA DIHs have the least variation, as 100 % of the participants offer technology services. This seems to be linked to the core competences of the DIHs, which emphasize the technical service categories.

According to our analysis, DIH facilities are one of the key aspects enabling the uptake of robotics for inspection and maintenance. The availability of facilities is linked to the testing and piloting services provided by the DIHs. The testing services can be provided regionally, however, the ecosystem of multiple DIHs enables companies to access specific test facilities on a broader scale.

3.2. The competition

Who are the competitors to the RIMA DIHs? Competitors are organizations owning technologies not available in the market and holding expertise similar to the expertise of the research organizations constituting the backbone of the RIMA Network. In that sense and in the perspective of being actors helping digitize European Industry at large scale, RIMA DIHs are rather exclusive as they gather some of the best research organizations in Europe. RIMA also aims to grow the network by adding other organizations that have the correct types of competence and focus. Asset owners doing their own R&D, or big service providers with internal R&D could possibly play a role close to ours. However, these organizations do not offer the independency that RIMA DIHs can offer. There are no real equivalent structures in Europe with the value proposition described above.

RIMA DIHs are new instruments designed to be long-lasting, renewing their technological offer regularly thanks to their involvement in research and development activities and offering their expertise through a set of services. A brief description of these services is provided in the next chapter together with some elements of the business model intended to make these services sustainable.

SECTION B

RIMA NETWORK PRINCIPLES

CHAPTER 4

SERVICES AND DIH ACTIVITY FIGURES

AKSEL TRANSETH, CLAUDIO PALIOTTA AND MARIANN MERZ

4.1. Overview of services

This section provides an overview of best practices for DIH services and explains how we have shared these best practices both within RIMA and via our support for building new DIHs in Europe.

DIHs generally provide a wide range of services to companies and public entities, and it is important that the companies and public entities that need services properly understand which services are available. To this end, the services can be divided into categories such as those in Table 1. Best practices for delivering services within each of the service categories are described later, in section 4.1. The categorization system divides services into 3 main categories and 14 subcategories as in Table 1 below. The Dutch RTO TNO named these 3 top-level and 14 next-level categories, but the description of the best practices within each category comes from all the partners in the RIMA project.

Ecosystem services	Short description
Community building	Create awareness of I&M robotics and regional ecosystem building through dissemination activities, workshops, 1-to-1 follow-up, matchmaking both within the DIH and to external stakeholders (e.g., industry associations, end users), and interaction with relevant national and international ecosystems.
Strategy development	Market research, market awareness, market/technology trends, mentoring by technical and application experts.
Ecosystem learning	Facilitate knowledge sharing by organizing informative seminars, workshops, conferences, field trips, newsletters, etc and distributing invitations, also other forms of information sharing.
Representation, promotion	Advertise and promote the capabilities and technologies of the RIMA Network and its members to suppliers, end users, etc. through roadshows, exhibitions, newsletters, etc.
Technology services	**Short description**
Strategic RDI	Long-term (2-10-year) R&D collaboration in combination with shorter-term spin-out commercialization projects (e.g., contract research), typically up to TRL 5.
Contract research	Shorter-term (0-4-year) R&D often compensating for short-term lack of resources or competences. Close customer-R&D-supplier interaction, technology demonstrators. Typically from TRL 5.
Technical support	Proof of concept, prototyping, commercialisation/industrialisation, small series production. Typically from TRL 6.
Provision of technological infrastructure	Access to lab facilities, equipment for lease/use (sensors, robots, robot tooling, etc.), rapid prototyping, 3D printing.
Testing and validation	Certification, product demonstration & testing, comparisons between products, help with safety assessments.

Business services	Short description
Incubator/accelerator support	IP strategy, IPR, business development, free and discounted services through vouchers and time-limited support.
Access to finance	Support for writing research grant applications, information about and connection to funding sources (e.g., open calls, public funding, investors).
Skills and education	General and topic-specific courses, hands-on training, mentoring and education, and general advice and consultancy.
Project development	Identification and concretization of project opportunities within R&D&I, consortia creation, proposal development and contract negotiation.
Offering housing	Offers of office space and workspace to spin-offs, start-ups and others – including co-location with DIH partners.

Table 1: Overview of service categories

An initial survey was conducted for the 13 DIHs participating in the RIMA project with the objective of mapping the number of services already on offer within each service category along with an estimate of the maturity level of these services. This data was used as a baseline of the Key Performance Indicators (KPIs) used by RIMA. New estimates were collected every 6 months for the duration of the project. Figure 12 and Figure 13 summarize the increase in the number of services and service maturity that was observed. While some of these developments may have occurred independently of RIMA, project participants indicate that much of the gain is due to the service-oriented focus in RIMA along with the process of organizing and supporting Financial Support to Third Parties (FSTP) winners within the project. Finding the right balance for the type and number of services on offer and finding a successful funding scheme for these services will require some experimentation on the part of each DIH, as they will be a function of the needs and local conditions of their region.

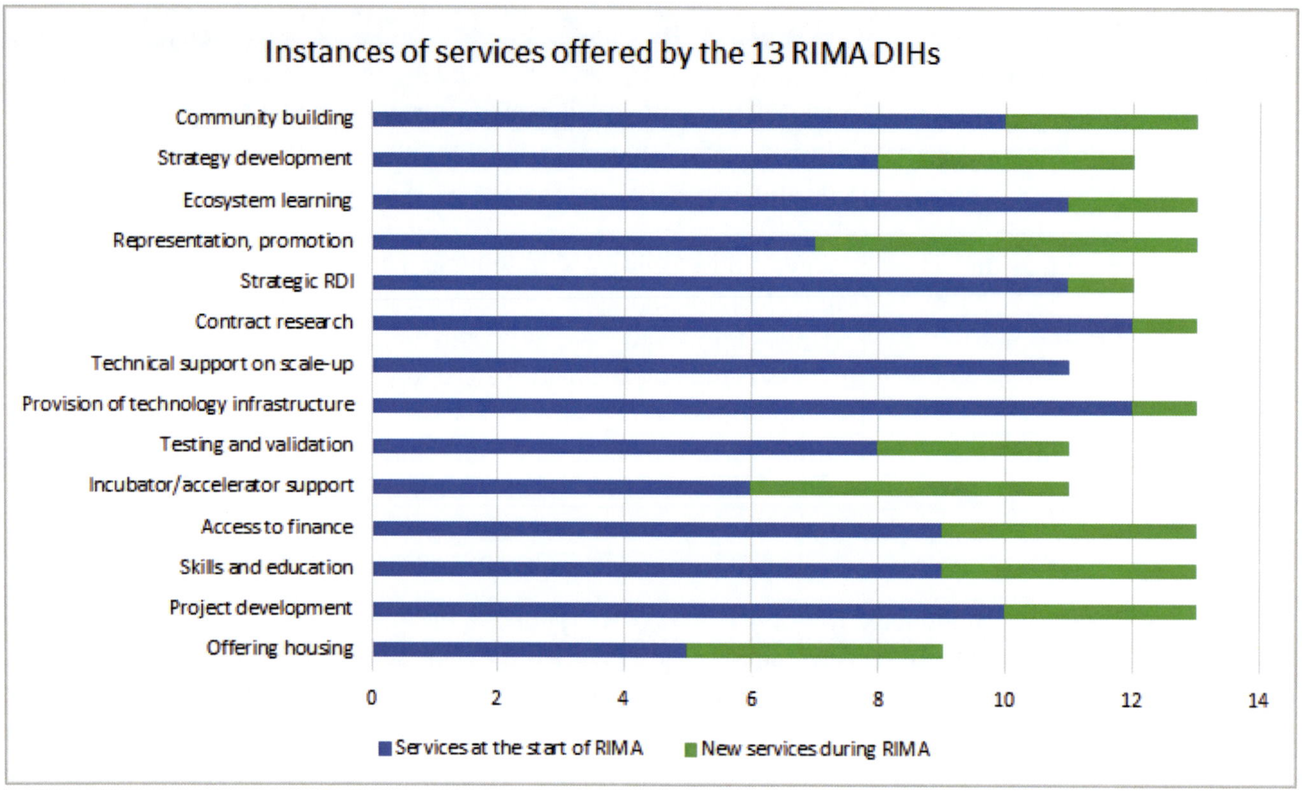

Figure 12: The number of DIH services on offer at the start and end of the RIMA project (for 13 DIHs)

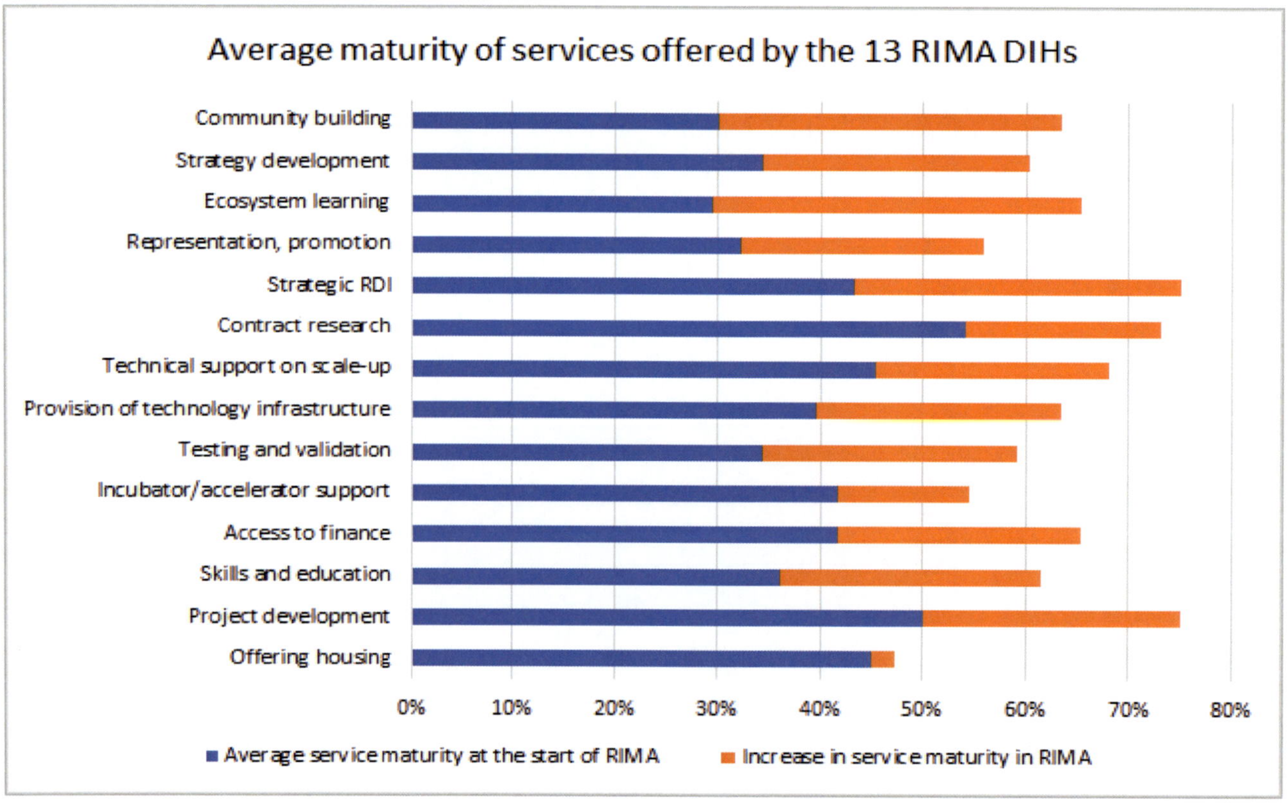

Figure 13: Service maturity development for the 13 RIMA DIHs during the RIMA project

SHARING BEST PRACTICES

The high-level picture obtained from the initial survey that mapped the number and maturity of services was used to help plan and focus the work in the RIMA Network activities. Based on the survey answers, community-building meetings were organized between small groups of 2-4 DIHs that could benefit from exchanging ideas and learning from each other in an informal manner. This practice was inspired by the good experience reported after similar activities in the DIH network DIH2. Based on the participating DIHs' service portfolios, relevant discussion topics were selected and distributed to the participants ahead of time to allow them to prepare. As the project progressed, the DIHs also shared their ongoing efforts to establish new services and the experience they gained from involvement in the FSTP work.

A lecture was developed for each main service category (Ecosystem, Technology, Business) discussing best practices and including specific examples from the RIMA Network. The lectures were made available online through the RIMA Community Portal. Their purpose was to provide inspiration for DIHs looking to expand their service portfolio or to improve specific services—or as "go-to" point for new DIHs. The next three subsections provide more details about the different services and summarize some of the best practices from the RIMA partners related to each service type.

4.2. Ecosystem services

The ecosystem service categories listed in Table 1 aim to create regional, inter-regional, national and international environments where SMEs, larger industries, research centers, system integrators and technology transfer centers work together sharing knowledge and experience and establishing fruitful collaborations. In such an environment it becomes important and relevant also to create market and technological awareness about establishing collaboration and cooperation among research centers and industry to strengthen the whole ecosystem and especially the industrial environment.

4.2.1 | Community building

The community building service is fundamental to:

- connect SMEs in the region to relevant actors within the value chain both within and outside the region;
- create awareness of the benefits of using I&M robotics;
- establish thriving ecosystems at regional, national, and international levels. This is to support SMEs in delivering their services and technologies regionally, nationally, and internationally.

Key aspects of community building include:

- Creating awareness of the opportunities offered by I&M robotics and digitalization.
- Facilitating participation of actors in the regional ecosystem.
- Connecting the organizations within a region and establishing strategic collaborations with organizations in other regions.
- Connecting the actors with relevant national, European and international industry associations.

Example

SINTEF has organized I&M robotics innovation workshops in Norway in collaboration with the Norwegian Society of Maintenance and the national unmanned aerial vehicle organization (UAS Norway). The purpose of the innovation workshops was to connect suppliers and end users within I&M robotics. The workshops were multi-domain with focus on I&M robotics in transportation (rail, road, aviation) and petroleum (subsea and topside). The program for the workshops was generally divided as follows: 1) end users share concrete needs within their industry, 2) suppliers (typically SMEs) give 3-minute pitches about their offerings, 3) short presentations on funding opportunities (e.g., EU, national, regional), and 4) an interactive session where end users and suppliers discussed concrete opportunities for collaboration. Participants gave these workshops very positive feedback.

4.2.2 | Strategy development

This service provides support to industrial sectors regarding market awareness and future directions of the market. Generally, this service is provided through the organization of workshops which gather experts in the field, followed by one-to-one follow-up with attending companies. Generally, such a workshop is attended by technical experts that work with technology development for the sector and market experts that are familiar with the current and future challenges of the field. The workshops are a good opportunity to share knowledge about the future of the market (5-10-year horizon) and about technical solutions for the current and future challenges. Furthermore, these events facilitate the meeting of technical and market experts who may not easily meet otherwise.

In addition to the workshops, one-to-one dialogues between DIH representatives and relevant companies were held to identify opportunities in I&M robotics for a specific company and to determine how to realize these opportunities commercially. Such a dialogue can be structured according to three services within strategy development as follows (see also Figure 14):

Evaluating automation opportunities

In-depth discussions with the company to identify their current need for I&M and identify relevant areas that could benefit from the introduction or increased use of robotics and automation. Typically performed by visiting a company (e.g., an end user) and getting them to talk us through relevant operations.

Application mentoring

Support with business modelling, fund raising and corporate engagement—and also determining whether there is an actual business need/case. This requires the mentor to be a senior expert, preferably with a wide background in business modelling and fund-raising and an overall vision of all the market domains addressed by the company using this service.

Technology mentoring

Matching the application areas identified by the evaluation of opportunities service in more detail with what is technologically feasible and suggesting relevant technologies and solutions. Technology mentors need a broad overview of I&M robotics technologies and solutions, including their limitations and possibilities.

These three services can be supplied by a single mentor or several, depending on factors such as how in-depth the work should be and the skills of the individuals involved.

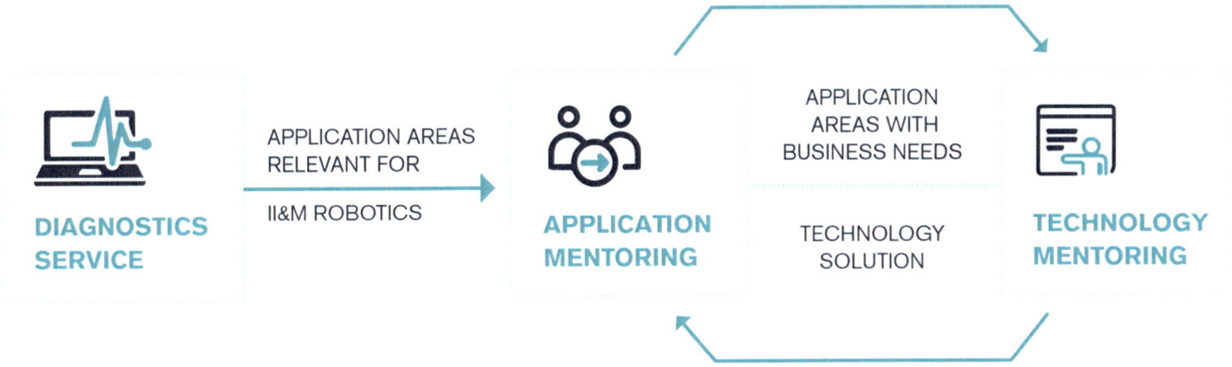

Figure 14: A combination of services within strategy development to increase uptake of I&M robotics

4.2.3 | Ecosystem learning

Ecosystem learning focuses on increasing the knowledge within a DIH about anything relevant for I&M robotics. Such knowledge includes end-user needs, benefits and risks with new technologies, funding opportunities, business development, and many more. This service is closely connected to community building services (since workshops within community building often contain some component of sharing knowledge) and skills and education services (as courses within education can be directly or indirectly used to share knowledge within the DIH).

The range of activities and events associated with ecosystem learning include:

- Seminars, workshops and conferences: about specific topics or more general.
- Trips with actors from the regional ecosystem, to visit relevant labs, clusters, trade fairs, etc. and learn from these.
- Newsletters and other sharing of relevant information.

The DIH can also actively inform interested parties about key outcomes from relevant conferences and work, new technology / business developments, and other relevant information about I&M robotics. This enables recipients to stay updated even if they are unable to attend the events or monitor all the relevant information channels.

Example

Water Europe organizes I&M robotics workshops for businesses, organizations, and local authorities. These meetings facilitate discussion about the latest available technologies on offer and the challenges that the sector faces. People can also discuss collaboration and initiate new projects. Typically, speakers give presentations in the mornings while the afternoon is dedicated to networking: business discussions and new collaboration opportunities.

4.2.4 | Representation and promotion

This service encompasses dissemination activities aiming at advertising and promoting DIH activities. This promotion is an essential activity for the sustainability of the DIH itself. In fact, promotion activities are paramount to giving visibility to the DIH and attracting new partners. One way of achieving this is to attend workshops or conferences with the aim of presenting and promoting the projects and activities carried out by the DIH. More specifically, it may be possible to attend technical conferences and workshops to share and discuss technical results and achievements obtained through the DIH work.

Roadshows and exhibitions to showcase new I&M robotic technologies and services are proven ways for developers and suppliers to reach out to potential customers. The roadshow can serve several important purposes including building trust in new technologies, an arena for suppliers to advertise their products and services, or a way to get feedback from potential customers on new features and systems. Furthermore, the roadshows enable e.g. asset owner personnel to get live demonstrations of new technologies.

Visits outside the DIH region are also considered to be promotional, but their high ecosystem learning value means they are described there, in section 4.2.3 above.

Example

SPRINT Robotics organizes roadshows for companies to showcase new I&M robotic technologies and services, mainly so that asset owners can learn about the new opportunities offered by these technologies and services. An important part of the roadshows is to include live demonstrations of I&M robotics—which gives the logistics challenge that the robots must also be transported between locations. Moving the roadshow between facilities enables key personnel that normally would not attend (due e.g. to time constraints) to see the exhibition. The suppliers are responsible for their own travel and accommodation and pay a participation fee to SPRINT to cover some of the costs in connection with the roadshow.

4.3. Technology services

The technology service categories are listed in Table 1. They aim to support technological development in different sectors (e.g., industry sector, public sector) related to I&M robotics. The support may be given in different forms and with different rules which are affected by local or national regulations and habits. Generally, the DIHs provide R&D services, support and infrastructure to carry out projects. Funding sources for these projects vary depending on their type (e.g., joint industry projects, regional, national, European, international, etc.).

4.3.1 | Strategic research, development, and innovation (RDI)

By strategic research, development, and innovation we mean long-term activities which typically aim to mature ideas and technologies originating from good results at low Technology Readiness Level (TRL). Strategic RDI activities for SMEs and larger companies in a DIH is typically within one or more of the technology areas described in Figure 15.

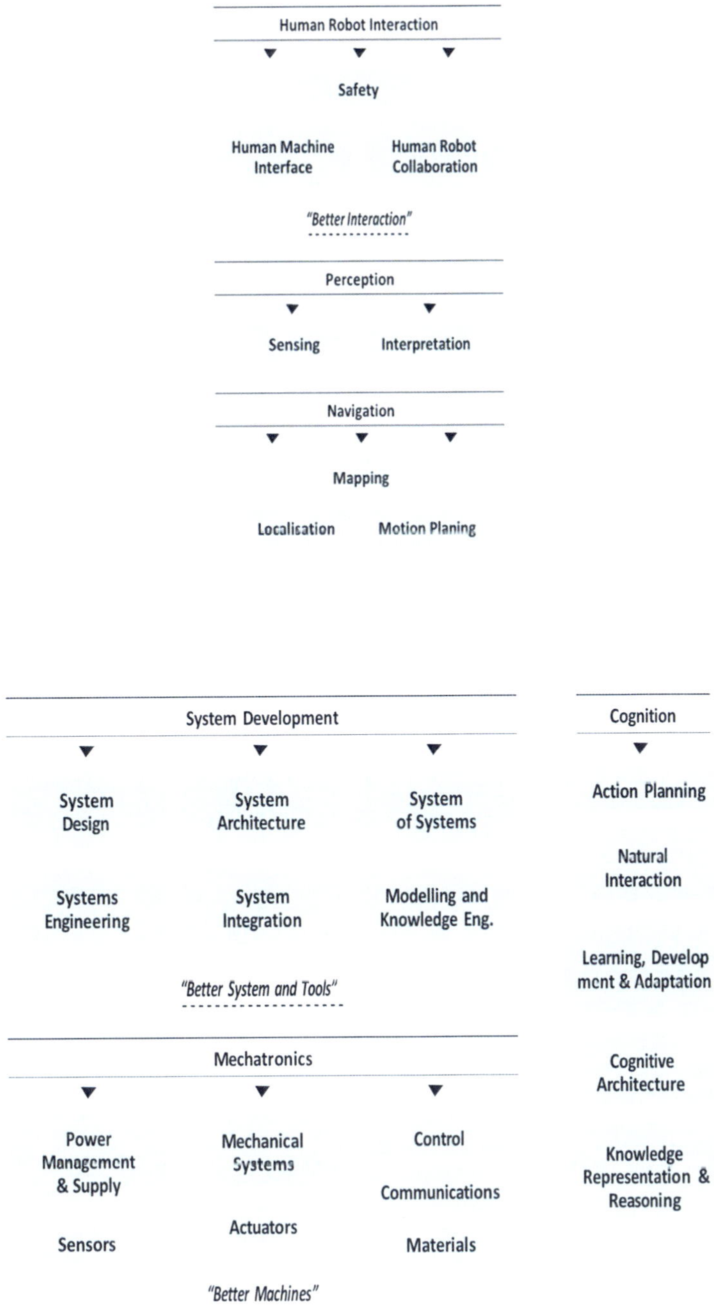

Figure 15: Technology clusters and areas from the SPARC SRA 2014–2020[1] page 68.

Further information about technology areas can also be found in, e.g., the euRobotics Multi-Annual Roadmap[2].

Strategic RDI contributes to keeping companies competitive in the long-term and thus supports their long-term strategies. Strategic RDI can also be a way of realizing long-term collaborations between companies and between companies and RTOs. Moreover, these kinds of RDI activities can be beneficial for creating visibility (through, e.g., media) for DIH activities, as plans for technology leaps often create more attention than incremental developments (which are usually the aim in shorter-term RDI initiatives).

Strategic RDI activities typically have a timespan of 5-10 years before the main results are materialized into commercial products. However, it can be advantageous to combine long-term RDI

1 The Partnership for Robotics in Europe (SPARK) "Strategic Research Agenda (SRA) for Robotics in Europe 2014 – 2020", https://www.eu-robotics.net/cms/upload/topic_groups/SRA2020_SPARC.pdf
2 https://www.eu-robotics.net/cms/upload/topic_groups/H2020_Robotics_Multi-Annual_Roadmap_ICT-2017B.pdf

with shorter-term spin-out projects as "branches" from the long-term efforts so that companies can continuously capitalize commercially from the ongoing effort. It is important for a DIH to have strategic RDI activities to remain in the technological forefront with respect to the state of the art within the strategic focus areas of the DIH. This enables the DIH to better support SMEs, mid-caps, etc. in their region.

Example

The Centre of Excellence SFF Autonomous Marine Operations and Systems (AMOS) is an eight-year strategic RDI effort organized and led by the Norwegian University of Science and Technology (NTNU). AMOS represents a close cooperation between large-scale asset owners, technology / service providers, a university (NTNU) and an RTO (SINTEF) and has led to several spin-off companies. The center has close collaboration with other national and international universities. The center is financed from funds from the Norwegian Research Council (a public entity which funds RDI in Norway), the industrial partners and in-kind contributions from the RTO and the university.

4.3.2 | Contract research

Contract research is often the primary research service for a DIH and is of key importance to SMEs, larger companies, and public entities wanting to solve challenges and/or gain new knowledge which is not already available in the market. Contract research is typically offered by a DIH to a company in cases where the company lacks in-depth knowledge about an area that needs development, and/or where the company needs additional resources, e.g. test facilities, for a limited amount of time in connection with R&D activities. The contract research can be in the context of a research project or as consultancy-like services in direct cooperation with the company. Contract research as a service is often targeted towards medium to large companies, government agencies, or institutes – and the projects typically span from six month up to 4 years in duration. Contract research with SMEs may typically involve some type of third-party funding such as national or European funds. Traditional contract research projects may be too slow-moving (in terms of time needed to set up the project and carry it out) for start-ups as they typically need results at a relatively high pace. Nevertheless, start-ups can also benefit significantly from traditional contract research projects.

Other DIHs, such as DFKI, may do contract research research in a different way, since national or European funding is involved, it is a funded project. Contract research in their case means that a company directly interacts with the DIH organisation and pays them with their own company funds.

System integrator companies may also be involved in paths to commercialization / industrialization of the results from contract research – normally towards the end of the project or just after. This is discussed further in the next service on technical support.

4.3.3 | Technical support

This service category covers mostly prototyping and small series production within I&M robotics. These can in many cases be seen in connection with contract research services: testing and validation and provision of technological infrastructure. To this end, a project with contract research can result in a prototype which is tested and validated using infrastructure provided fully or in part by the DIH. The reasoning behind the prototype and its purpose affects the development strategy and may significantly influence the budget required as well as the type of competences that are needed for the development work:

- Demonstrate and test a concept: Development and/or assessment of a prototype which demonstrates a concept but does not necessarily reflect how the final system will be implemented in terms of choice of, e.g., materials, programming language, user interface.

- Support designing automated production systems, for the scale-up step from small, manual "garage" production to small-scale automated production. Including recommendations for small changes to the product to make this automation easier.
- Final steps towards a commercial/industrialized product: Development of a prototype which closely resembles (in terms of choice of materials, programming language, etc.) what is believed to be the final product/service.

Finance for prototyping activities may be organized in connection with contract research or as stand-alone paid by the company or, sometimes, public funds. National funds in some countries are targeted towards prototyping and demonstration activities, and some FSTP projects, e.g. those within RIMA, can also support such activities financially.

4.3.4 | Provision of technological infrastructure

Technological infrastructure generally encompasses lab facilities, system platforms, test equipment, office facilities or more generic equipment. The provision of technological infrastructure is key to the "test before invest" mantra of DIHs and may be facilitated through several different approaches:

- The infrastructure may be made available for a project that the DIH participates in and then DIH personnel operate the lab and the equipment according to scope and indications defined in the project.
- Facilities may be rented to industries using different forms of leasing agreements. That is, labs and equipment may be rented out directly to industries and personnel support may be included or not. In this case, the costs are charged directly to the industries.
- Infrastructure may be rented out using a voucher system where the vouchers are given to companies as a funding form from regional or national development plans.
- Free use of infrastructure granted to promising start-ups as a part of a development and support plan for them. The DIH indirectly supports the start-up by covering the infrastructure costs and when and if the start-up is mature enough to sustain itself, a reasonable price for equipment rental may be negotiated. In this case, the DIH gambles on the fact that a good return on the investment is generated by future income from research contracts obtained with and/or through the developing start-up.

4.3.5 | Testing and validation

This service gives support to companies concerning certification, product qualifications and product demonstration and is closely linked to the service of provision of technological infrastructure. The DIH may offer support in the form of testing facilities and know-how to prepare the product and its associated documentation for a final certification which may be given by qualified external associations. Some DIHs perform certification of certain types of products themselves, but in general it is more common that DIHs only support the development of a new product towards this objective. It is then clear that the DIH may also support product demonstration to understand the product's current status, or the DIH can facilitate demonstration to interested third party companies or potential clients. This service could be financially supported by rental fees for the DIH facilities and by hourly rates for necessary personnel support.

4.4. Business services

Business-related services encompass key aspects of developing projects and supporting them on the way to commercialization. Skills and education in terms of both technical and business knowledge provide the foundation, categorized into, e.g., project development, access to finance, incubator / accelerator support and offering housing. Business services focus on creating activities that can turn knowledge into market-ready products and services.

4.4.1 | Incubator/accelerator support for start-ups

Some DIH members host incubator/accelerator environments that support start-ups. Support services include helping start-ups understanding different strategies regarding Intellectual Property Rights (IPR), discover start-up funding, and discuss market focus. DIHs can provide input on market needs as the DIH member organisations probably have close connections with potential customers. Other RIMA services can also support start-ups, such as the ecosystem subcategories community building and ecosystem learning, or the technology subcategories contract research and provision of technology, as described in different sections of this book.

Some incubators offer support regarding IPR strategies and management, for example helping to set up a start-up's strategy for how to secure and produce value from the IP in the company. A patenting approach can be one way of achieving this, but it is not the only way. DIHs can provide support with e.g., writing funding applications. Successful applications result in projects that deepen cooperation with the other project partners, who normally have similar or complementry interests. Bilateral RTO-company research projects without external funding from, e.g., regional/national/EU grants, may in many cases be too costly for a start-up.

4.4.2 | Access to finance

Services that help companies and public organisations to access finance rely heavily on our knowledge of funding sources and our networks of contacts. These services include: information about the many funding opportunities available, discussion of the focus and merits of each one, help with writing applications, and help with building winning consortia. Many grants are competitive, whether from regional/national funding, from the EU, or smaller amounts through private foundations. DIHs can also help find partners (SMEs, larger companies, asset owners) to start Joint Industry Projects (JIPs) without public funding. It is healthy not to be dependent on a single finance source, to be able to survive changing market conditions.

4.4.3 | Skills and education

Research, industry, and academia need education to be relevant and up to date so personnel attain a skillset relevant for I&M robotics. Academic institutions need to stay up to date to attract students, and research institutes need to stay industrially relevant to attract funding. DIH services in skills and education can help sustain this co-dependency.

RIMA DIHs provide courses for the following important target groups within I&M robotics:

- University students, typically at bachelor, master and/or PhD level: These give the students the opportunity to seek a job within I&M robotics. In addition, they could get specific training to enable them to use and develop new I&M robotic technologies and services. In this way, RIMA helps prepare a skilled workforce.
- Technology developers: Teach them about the uses of new technologies that are available, for example, what types of AI can be integrated into their product or their production; or new ways of approaching quality control. Also provide guidance about relevant business development strategies.
- Service providers: Explain how they can use new I&M robotics efficiently and provide guidance on safety and relevant business development strategies.
- Asset managers: Inform them about their options for using I&M robotics to protect their infrastructure and how this can affect their business.

Example

An example of a service in this area is "The Teaching Factory" at the Laboratory for Manufacturing Systems & Automation (LMS, Greece) which is designed to increase collaboration between academia and industry. Industrial organisations provide relevant, real-life engineering challenges. The university provides teams of students that propose solutions to the challenges. Dialogue along the way fosters healthy idea generation and industry/academic collaboration. In their final year, students are selected on merit and, assisted by an LMS representative, begin to interface with engineers from the industry. Students and engineers meet periodically to discuss ideas, possible solutions, prototypes and potential final demos. This initiative gives good visibility to good students who may subsequently get hired—and, at the same time, may lead to innovative solutions to technical problems that industry is facing.

4.4.4 | Project development

Projects are an essential DIH activity and there can be a range of different projects that involve different types of stakeholders and financing options. DIHs can support SMEs, larger companies, and public entities through joint projects. Some of these are internal to a DIH (i.e. all partners come from the same DIH) while others also involve entities outside the DIH (e.g., from other regions). Project development starts by establishing the project, and the approach selected will be influenced by the type of project. Activities such as those within community building, ecosystem learning, and strategy development may be important activities for initial project development. A key to project development in many cases is to understand the needs of the stakeholders involved and how the ecosystems around the different stakeholders work.

DIHs are organized in different ways – also in terms of project development. At some DIHs there are business development groups responsible for project development, other DIHs have business developers employed within the individual research groups in the organizations. There can be benefits to establishing a team for each specific business development case and having close collaboration between business developers, technical experts, legal and project managers in such a team. Sometimes individuals can have several of these roles at once (e.g., a research scientist may also have expertise in business development). Business developers can formally leave a project initiative when/if it materializes into a project, but there can be benefits to keeping some involvement throughout the project to maintain relationships with the clients.

4.4.5 | Offering housing

This service offers infrastructure to small promising companies that need support in the initial phase of their activities. Among the multitude of ways this service may be offered, a large organisation within a DIH may offer office space, lab access and/or mentoring to start-ups at a competitive price or, even better, for free. This is often achieved via financial support from local or/and national governments. The selection process is normally competitive, carefully choosing the most promising companies to benefit from this support. Note that hosting several small companies in the same building may offer several advantages, for example in terms of synergies. That is, small companies or start-ups working in similar or totally different sectors may support each other via informal information exchange or more formal collaborations. They might end up starting new projects together, benefit from the opportunity to broaden their knowledge, they may just support each other during "corridor talks" or even with reciprocal consultancy support. In other words, a side effect of this service may the implicit establishment of a small ecosystem.

4.4.6 | Refining the service categories

It should be noted that defining detailed services within each service category was challenging due to several factors. First and foremost, many of the DIHs were relatively new and not complete with all relevant actors at the start of RIMA – they required time to mature and decide on overall strategies and priorities before organising detailed services. Second, the successful offering of specific sustainable services (such as the rental of a specific laboratory facility) requires a period of service development and refinement (including the development of new legal and liability contracts /

documentation) as well as extensive awareness and marketing efforts toward potential users. This broad-scale marketing was not possible within the RIMA project. Approximately mid-2021, the 13 RIMA DIHs were asked to define their applicable services and were encouraged to break down each of the 14 service categories into a finer grid of services. The results are summarized in Table 2.

RIMA SERVICE CATEGORIES (Based on TNO definitions)		REFINED SERVICE CATEGORIES	RIMA DIH INSTANCES
ECOSYSTEM	**Community building**	Brokerage, matchmaking	11
		Awareness creation	13
		Dissemination	12
		Ecosystem building	10
	Strategy development	Market research, market awareness, market/technology trends	8
		Technical consultancy	6
		Mentoring by technical and application experts	7
	Ecosystem learning	Resources and knowledge sharing	11
	Representation, promotion	DIH network promotion, promotion of the offer	11
TECHNICAL	**Strategic RDI**	Research and development collaboration	12
	Contract research	Innovation projects: technology demonstrators, proof of concepts	13
	Technical support	Proof of concept, prototyping, commercialization/industrialization, small series production	11
	Provision of technological infrastructure	Lab facilities, equipment for lease/use (sensors, robots, robot tooling, etc.), rapid prototyping	13
	Testing and validation	Certification, product demonstration & testing, comparison of products	7
BUSINESS	**Incubator/accelerator support**	Business development	9
		Funding opportunities investigation	3
	Access to finance	Connect to funding sources	8
	Skills and education	Hands-on training	12
	Project development	Research, development and innovation opportunities	12
	Offering housing	Providing spin-offs and start-ups with infrastructure	10

Table 2: Refined service categories proposed by RIMA DIHs along with service offerings in June 2021

It can be noted that several of the proposed services overlap with each other, e.g., the DIHs offer either "Technical consultancy" or "Mentoring by technical and application experts" under ecosystem services. In general, the refinement of service categories for a DIH must consider not only the specific needs of the local ecosystem, but also the geographical, legal, cultural, economic and political conditions of its targeted region/country and its focus industries.

4.4.7 | Value creation with new DIHs

The RIMA project gave careful consideration to the integration of new DIHs. While network expansion was seen as desirable in itself, added value must be foreseen both for the DIH and for the RIMA network. This means that the expansion should preferably be into new regions (i.e., not already covered by the RIMA network) and the DIH candidate must have a strong I&M robotics focus. Growth enabled through DIHs that contribute with shared value creation will lead to an upward spiral of potentially available value, meaning a win-win situation for all the stakeholders in the DIH ecosystem, as, in a non-exaustive list, system suppliers, system integrators, service providers and end users

Let us first consider the role of the DIH itself. Recent technological developments (e.g., deep learning, component miniaturization and cost reduction, progress in connectivity and high-speed communication) enable increasingly complex I&M robot technologies and applications. To achieve the full potential of I&M solutions in a time- and cost-efficient manner, the exchange of insights and information between the different types of stakeholders is critical. The DIH should be the driving force behind this collaboration by acting at a level beyond the individual stakeholder towards common industrial, regional, and societal goals.

To successfully fill this role, the DIH must be committed to frequent monitoring of its service offer and to making constant adaptations and improvements to services in order to continue to provide added value for its stakeholders in a changing market. While a DIH is often well aware of its main shortcomings, some of the gaps may be in a blind spot. Collaborating with a DIH network and sharing information about the organization of services and lessons learned can help a DIH to identify aspects of their operation with room for improvement. Furthermore, all the DIHs and associated stakeholders in the network can benefit from knowledge about, and potential access to, world-class research facilities and laboratories across Europe.

CHAPTER 5

BUSINESS MODELS

CHRISTOPHE LEROUX

5.1. RIMA Network business model

The purpose of this section is to explain the ability of DIHs to create value through their activities. An outline is given using a business model canvas that is filled in for RIMA in Table 3 below. The key elements are: key partners, key activities, key resources, value proposition, customer segments, customer relations, channels, cost structure and revenue streams. Key partners are mostly the other RIMA DIHs. Key activities are the services described in section 4 above. The value proposition of the RIMA Network was discussed in section 3 above. Details of cost structures and revenue streams relevant for RIMA follow below.

Details of business model canvas elements – especially activities—for each individual RIMA DIH are found in section 6 and the appendix.

BUSINESS MODEL CANVAS

Despite criticism of their reductive aspects, "Business Models" seem unavoidable to allow business organisations to be seen and evaluated as "value creating/defending systems". This section uses a business model canvas to represent in a synoptic the essential facets of the business model of a RIMA DIH. The business model canvas is a method that makes it easy to build a business model, by presenting the key aspects on a single page in the form of a table. The business canvas facilitates identification of the strengths and weaknesses of organizations and the value they create and for whom. These descriptions and definitions are important for the DIH business plans to be sustainable.

Several methods can be used to make a business model canvas. The best known and most widely used method was proposed by Alexander Osterwalder and Yves Pigneur[1]. We use their method in the rest of this document, because it is clear and simple.

1 Osterwalder, A., Pigneur, Y., Tucci, C.L. (2005). "Clarifying business models: Origins, present, and future of the concept". *Communications of the Association for Information Systems*. **16**: 1–25. doi:10.17705/1CAIS.01601 – via Association for Information Systems.

<table>
<tr><td colspan="2" rowspan="2">Business Model Canvas[2]</td><td>Designed for:</td><td>Designed by:</td><td>Date:</td><td>Version:</td></tr>
<tr><td>RIMA network</td><td>RIMA partners</td><td>5th April 2021</td><td>1.0</td></tr>
<tr><th>Key Partners</th><th>Key Activities</th><th>Value Propositions</th><th>Customer Relationships</th><th colspan="2">Customer Segments</th></tr>
<tr><td rowspan="3">Professional and sectorial associations, asset owners, asset operators.

Techno & service providers, system integrators, research organizations, certification bodies, regional clusters, public authorities.

Description of needs, constraints, capacities, offer.</td><td>Ecosystem services: Community building, strategy development, ecosystem learning, representation, promotion, etc.

Technology services: strategic RDI, contract research, technical support, testing and validation, etc.

Business services: Incubator & accelerator support, access to finance, skills and education, project development, housing, etc.</td><td rowspan="3">Enable & assist SMEs and other companies. Help tech developers to bring innovative technologies to the market. Help end users to automate their I&M.

The main value propositions are the following:

-Contribution to reducing time for product development and testing;

-Favour cross-fertilization between sectors and regions;

-Consolidation of technical market needs.</td><td>Techno-commercial activity.

Membership in associations, in technological platforms.

Communication, dissemination of results.</td><td colspan="2" rowspan="3">Asset owners, asset operators, techno and service providers involved in inspection and maintenance of infrastructure: especially in oil and gas, nuclear, water supply and sanitation, energy production & distribution, land transport, sea & air transport, communication.

Possible connection with other domains: public sector (municipal), construction, agriculture, healthcare, security.</td></tr>
<tr><th>Key Resources</th><th>Channels</th></tr>
<tr><td>Personnel with specific skills (e.g. business developers, communications personnel, technical experts). Laboratories, sandbox space, etc.

Networks of organizations with relevant skills and infrastructure.</td><td>Joint projects, conferences, matchmaking, brokerage, workshops, tech and info days, demonstrations, conferences, fairs, social media posts, videos</td></tr>
</table>

2 Designed by: The Business Model Foundry (www.businessmodelgeneration.com/canvas). Word implementation by: Neos Chronos Limited (https://neoschronos.com). License: CC BY-SA 3.0

Cost Structure	Revenue Streams
Personnel costs: management, experts, researchers, business developers, consultants, promoters, jurists, IP managers. Robotics equipment, test infrastructures, maintenance, consumables.	Bilateral contracts, tax refunds, technology transfer, licensing. European, national and other public funding for tech research, innovation, disruptive solutions and proof of concepts in competitive calls. More general public funding support for DIH activities and SME support. Internal (pays for tech research and disruptive solutions). Membership fees.

Table 3: Business model canvas for a DIH in RIMA

COST STRUCTURES

The business model canvas enabled the identification of customer segments for which different service packages can be designed. The experience of the more mature RIMA DIHs makes it possible for even new DIHs to make an estimate of the activities and resources required for each service offering as well as outlays and cost. A preliminary budget can then be established by comparing the estimate with the total resources available and identifying the time and the additional resources required. Different customers have different needs, but by establishing a starting point in terms of a first offering and budget estimate, and applying the process described in Figure 16 iteratively, the original estimate and offer packages can be adjusted or changed after experience from a number of projects provides more data. This will ensure maximum value for the customers and maximum return on investment for the DIH. The DIH needs sufficient income from the customer to make the service (and the DIH members) sustainable. It is important that internal management processes and communication channels are in place within the DIH organisation, so that each individual (e.g., business developers, engineers, researchers) understands the activities required to deliver the various offer packages.

SALES & MARKETING

Sales and marketing efforts are critical in establishing a sustainable business and require dedicated resources. In RIMA, DIHs divide the sales and marketing effort into three phases as illustrated in Figure 16:

- engagement with potential customers and listening to their desires,
- elaboration of a techno commercial proposal, and
- contracting.

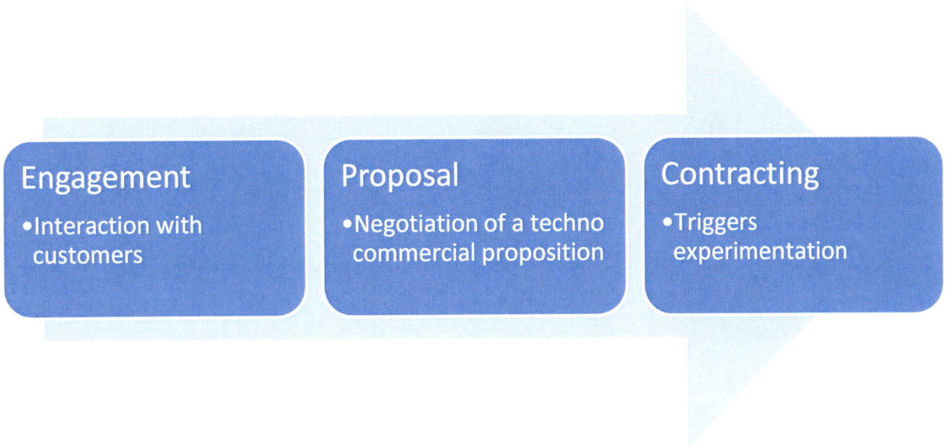

Figure 16: Breakdown of sales and marketing effort in RIMA

Given that the time between first contact and delivered service can be long, up to several years, it is necessary to build an extensive pipeline of potential customers. Only a reduced percentage of these leads can be translated into actual technology transfer or a contract.

The experience of the RIMA DIHs enables estimation of the proportion of qualified leads that might result in a project, in relation to the number of first contacts. Based on this figure, the resource and time requirements in relation to the current pipeline can be assessed and the budget adjusted accordingly. RIMA DIHs use customer relationship management tools for this exercise.

Sales goals are agreed by each DIH organisation and specific targets are set based on current estimates of organisational capacity. Target objectives for RIMA have been defined in D1.1[3]. The sales efforts are continuously followed up. It is important to hold discussions with customers to gain an understanding of their perspectives, as this is key to the sales effort and a successful final result.

There are many approaches to understanding the needs of potential customers as a basis for establishing contracts. Such approaches include, e.g., "design thinking"[4,5] as well as more task-specific techniques such as the "5 whys"[6] for understanding the root cause of a potential customer's challenges and the SPIN[7] methodology – a targeted interview technique described below which can be used to elicit information about a customer's specific challenges. Since its publication in 1988, SPIN selling has become one of the most popular guides for B2B sales worldwide.

SPIN selling involves:

- Situation questions, which collect facts about the customer's existing situation. These should be formulated in such a way as to gain a personal understanding of the customer.
- Problem questions, which ask about the customer's 'pain' and focus the customer on clarifying and detailing their specific problem. These elicit the customer's Implied Needs.
- Implication questions, which ask about the effects of the problem, before talking about solutions. These questions help the customer to explicitly recognise the seriousness of the problem and increases the customer's motivation to find solutions.
- Need-payoff questions, which probe for Explicit Needs and encourage the customer to tell you about how your solutions will meet these needs, rather than leaving you to explain the benefits to the customer. This has a more positive impact on the customer and is perceived as much less 'pushy' as a sales technique.

3 10 engagements per month (KPI3.10), 10 to 15 proposals per year (KPI3.11) and 5 to 10 contracts per year for the whole consortium.

4 Lewrick, Michael, Patrick Link, Larry Leifer. *The Design Thinking Playbook*. Hoboken, NJ: Wiley, 2018

5 https://en.wikipedia.org/wiki/Design_thinking

6 https://en.wikipedia.org/wiki/Five_whys

7 Neil Rackham, *SPIN Selling: Situation Problem Implication Need-payoff*, May 22nd 1988 by McGraw-Hill Education

FINANCING, REVENUE STREAMS

The revenue streams of a DIH describe how it makes income within each customer segment and from its different services. The offering to each customer needs to be matched with a suitable financing instrument, either directly (e.g., through a bilateral contract), or indirectly (e.g., though membership fees, future contracts, etc.).

In the overall business model canvas for RIMA DIHs in Table 3 above, we outline a set of revenue streams. These can usually be divided into the following categories:

- Free for certain customer segments (but may still in some cases generate direct or indirect revenue for the DIH).
- Recurring revenues (e.g., membership fees).
- One-time / limited-time fees (e.g., bilateral contracts or publicly-funded projects).

In some cases (e.g., for some ecosystem services) DIHs use a "free" business model where at least one customer segment receives the services for free. In some cases, such free-of-charge services are paid from elsewhere so can still create direct revenue for a DIH. For example, public funding bodies often support start-ups and SMEs by paying for ecosystem-type services (conferences, workshops), or business and technical support services. Their aim is to initiate innovation activities among SMEs, large companies, and public organizations. Other examples of DIHs using free business models may be within project development and access to finance, e.g., helping to establish projects with industry or public organizations. In this way, DIH members can provide free support with the potential (future) economic upside of receiving income should, e.g., project proposals be successful.

It is however doubtful that general services against direct payment will ever work for small or micro-companies. It can work if the result of the service will be a tangible asset for the SME: a certification or a specific test for instance. Some services (especially networking services) will always be free since it is almost impossible to establish the value of a contact in advance. Other services (especially those related to obtaining funding) will probably only be viable if the "no cure, no pay" (contingent fee) principle is followed.

One-time / limited-time fees typically include contracts between a DIH and one or several industries and/or public organizations, and makes up a very important part of the revenue of a DIH.

The challenge of finding appropriate funding for a customer project requires the same careful handling as converting a potential lead into a new customer. DIHs in RIMA do not have the financial resources to fund all of the services shown above themselves, so complementary funding should be found.

The list below presents some solutions that have been implemented and possible alternatives by the RIMA DIHs to financially underwrite their service activity.

1. **Large enterprises** normally have the financial resources to engage in development projects, either through direct financing or through part-public funding. Maintaining competitiveness requires that the industry invest in digitalisation of the entire supply chain, of which robotics and automation form a subset. This is an opportunity for RIMA DIHs.
2. **Regional, national, and European bodies** have funding instruments to support RDI and to support SMEs. Some aspects remain constant, e.g., the conditions for the National Research Council competitive grants always require that the proposals have a strong link between business potential, innovation potential and significant R&D challenges. However, the topics that the funding agencies wish to focus on and what they specifically wish to achieve with their money keep changing depending on the percieved need. Funding instruments and the modalities of access are also frequently updated—as illustrated by the recent NextGenerationEU

recovery plan for Europe[8]. Therefore no details can be given here, as they would be out of date before this book could be published.

3. Some **banks** are interested in enabling their customers' growth by financing development projects. Many private foundations exist that pay for specific types of work, but normally in small amounts.
4. The growth potential from investment in robotics and automation is also interesting for **private investors and business angels**, if they feel they understand enough about how to assess the risks.
5. **Membership fees** or per-service fees for access to information are used by some DIHs.
6. **Licensing** is also used to get some income from innovations.

5.2. Services business model and business plans

This section presents the business plan and business model used by RIMA DIHs to ensure service sustainability. For each service, we provide a brief recap of the offer: what, how, who, cost, and involvement.

DIHs involved in the RIMA network all have experience in interacting with and delivering services to industry. The services are offered following the iterative agile process illustrated Figure 17 to take advantage of the lessons learned and create an optimized offer.

RIMA DIHs also learn from each other to refine the services offered.

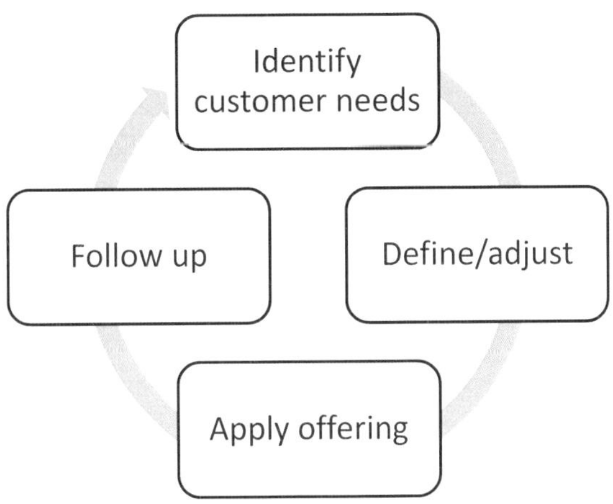

Figure 17: Iterative service design process for continuous improvement of the organisation's offering

8 https://ec.europa.eu/info/strategy/recovery-plan-europe_en

CHAPTER 6

RIMA NETWORK OF DIHS

JONATHAN VAN DER MEER AND JUAN ANTONIO PAVÓN LOSADA

In this section there is a description of each RIMA DIH and a resumé of the key services that it offers. More details of these services provided by each DIH are given in the Appendix.

ROBOTICS DIH SPAIN:

GRVC - Robotics, Vision and Control Research Laboratory. Universidad de Sevilla.

Website	https://grvc.us.es/
Email	grvc.us.es@gmail.com
Phone	+34 954 48 67 48
Address	ETSI – Escuela Técnica Superior de Ingeniería. Camino de los Descubrimientos, 41092, Sevilla. Spain.

ABOUT

The **Team GRVC Robotics Laboratory** at the University of Seville leads the **Robotics DIH Spain** with the active participation of **FADA CATEC** and the support of **Technological Corporation of Andalusia (CTA)**. These three entities are currently committed to contributing to the digitalization of industry and services through robotics. The DIH offers services that help accelerate the development and deployment of robotic technologies in different sectors, integrating different actors into the value chain, such as research and technological development institutions, system integrators and end-users.

DIH INFORMATION

THE ROBOTICS, VISION AND CONTROL LABORATORY GRVC

The Robotics, Vision and Control Laboratory GRVC[1] is a large laboratory (> 75 members) which plays a relevant international role in robotics, and particularly in aerial robotics and unmanned aerial systems.

GRVC combines scientific research excellence with technological development and transfer to companies. I&M activities benefit greatly from excellent collaboration with the asset owners to better understand end-user needs.

1 https://grvc.us.es/

Figure 18: GRVC focused on aerial I&M applications

Such collaborative initiatives enable the GRVC to focus on generating new technologies and knowledge while transforming the I&M activities in sectors like oil & gas, roads and civil infrastructures, transport hubs, etc.

GRVC offers world-leading **laboratory and test facilities** in the field of aerial robotics, which highly benefits the demonstrative activities that many SMEs and end users require from us on the effective use of robotics for risk avoidance or improved efficiency within their I&M processes.

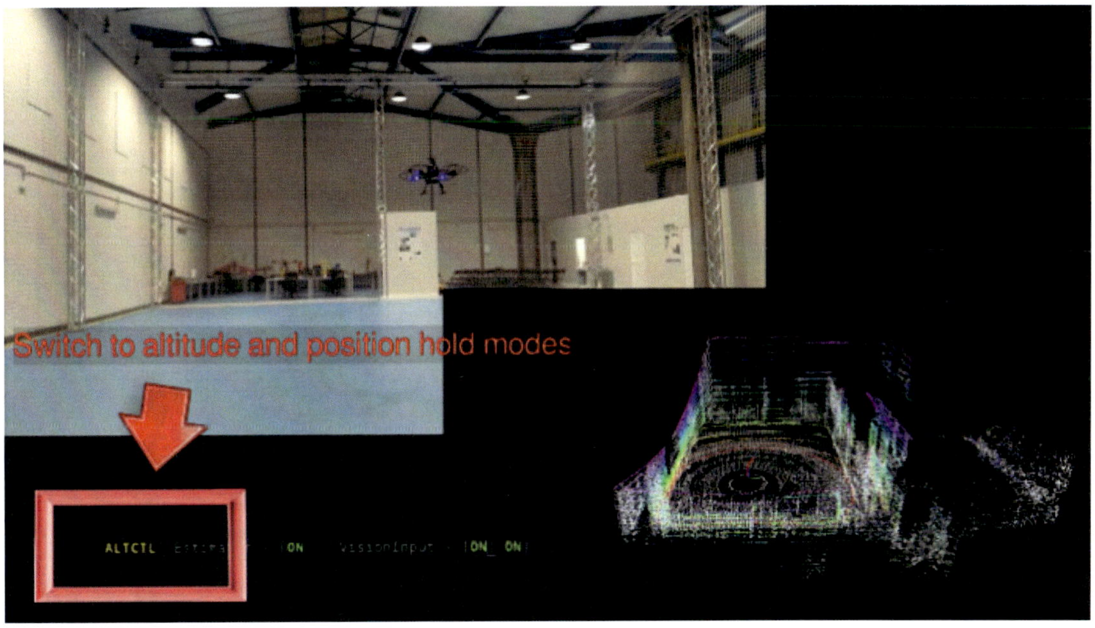

Figure 19: Indoor facilities at GRVC. OPTItrack as a ground-truth for refining navigability features

Figure 20: Outdoor facility used for testing multi-UAV collaboration

As part of their mission, GRVC moves technologies from laboratory to real-life scenarios. The business rules are specific to each sector and generating viable concepts and products can only derive from knowing the intended deployment conditions of our robotics solutions. Figure 21 shows some of the proof-of-concepts developed in the last few years during a variety of European projects.

Figure 21: GRVC advances in aerial robotics.

CATEC

CATEC supports GRVC activity with its technological center in which they develops its activity based on seven strategic lines:

1. New composite materials, their production processes, life cycle and recycling. Automation of production, assembly, inspection, and testing.
2. New onboard systems: electrical, hydraulic, compressed air, fuel, etc. directed to achieve the most efficient and ecological plane.
3. New avionics systems and subsystems, especially with application to general aviation, sports, business, and Very Light Jets.
4. New techniques of air traffic management, including management of airports and their technical facilities.
5. Unmanned aircraft, applications, and associated technologies. Special attention to commercial, civil and security applications. UAVs as flight test platforms.
6. Development of computational simulation techniques and their application to systems, platforms and / or vehicles (integrated systems)
7. Development of techniques, components, subsystems, systems, and spatial applications.

CATEC has a staff of more than 60 researchers as well as specialized technical personnel, most of them engineers from industrial, aeronautical, telecommunications, materials, computer or chemical fields, and there are also some graduates in other technical subjects.

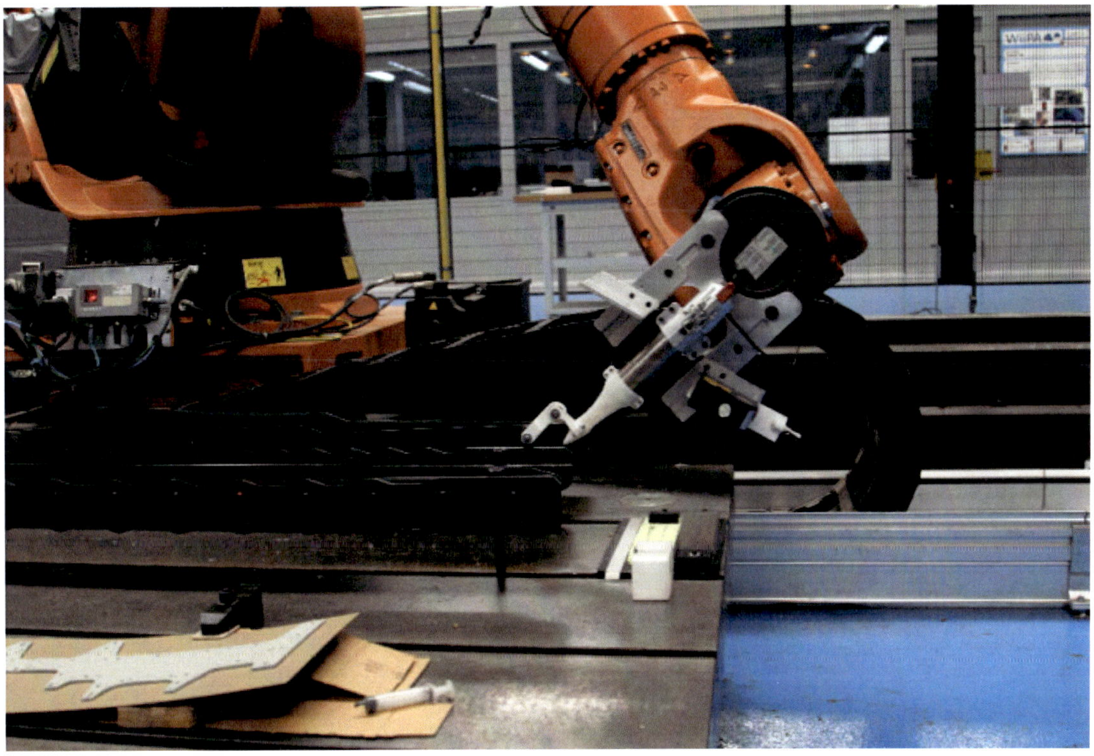

Figure 22: CATEC's capabilities on composite manufacturing

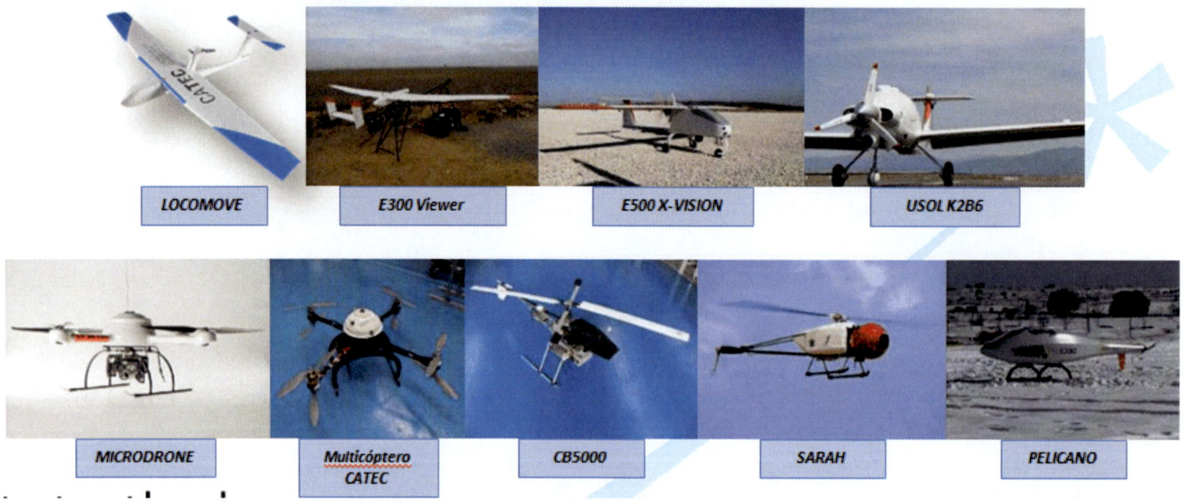

Figure 23: List of aircraft

Figure 24: An exampleof CATEC's experience in UAS/RPAS technology for I&M tasks

TECHNOLOGICAL CORPORATION OF ANDALUSIA (CTA)

CTA is organized as an innovation cluster which is open to companies throughout Andalusia, regardless of their size. The sectors that are mostly addressed in Robotics DIH Spain are: aerospace and production processes, agri-foods, biotechnology, construction and civil engineering, energy and the environment. CTA incorporates their considerable experience in promoting transverse projects which are based on these sectors.

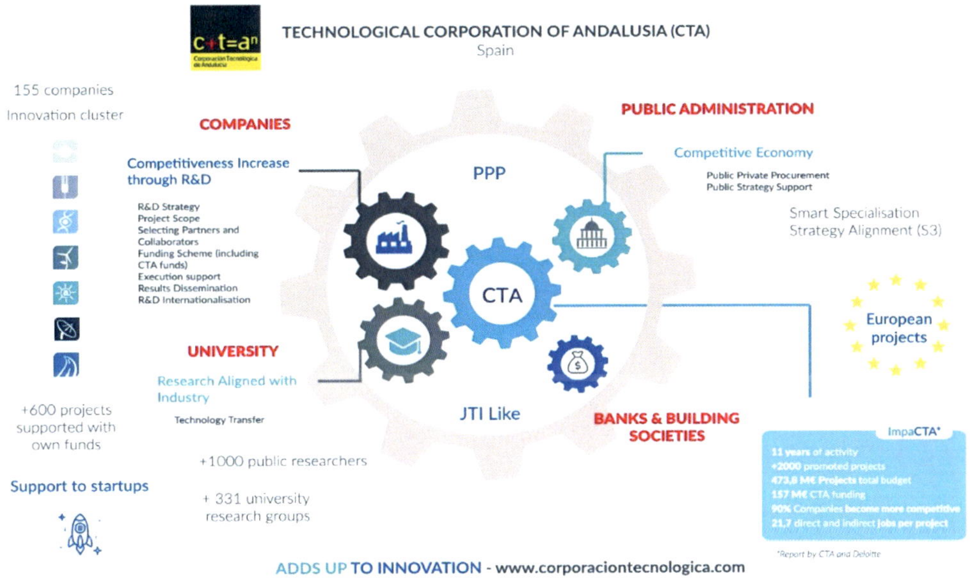

Figure 25: CTA as part of an innovation ecosystem

EXPERTISE AND SCOPE

Robotics DIH Spain includes a set of technologies and expertise in robotics and related technologies in a manner that is not available on the market. We offer a set of services to favour digitizing I&M of assets and infrastructures. While this background enables USE/CATEC to be very active in robotics research areas (e.g. in aerial robotics), they also aim at the effective technology transfer to the industrial sector. Hence, they are supporting the DIH services in creating a framework facilitating deployment of technologies, building connections and trust and understanding the needs of industrialists.

The geographical scope is Andalusia region first and extending into the national scene. It extends also into Europe, although typically from contacts harvested during previous collaborations in EU programme work.

SERVICES

SERVICES RESUME for the Robotics DIH Spain
Business Services: Access to finance; Incubator/accelerator support; Offering housing; Project development; Skills and education.
Ecosystem Services: Community building; Ecosystem learning; Representation, Promotion; Strategy development.
Technology Services: Contract research; Provision of technology; infrastructure; Strategic RDI; Technical support; Testing and validation.

CONSORZIO C.R.E.A.T.E

CONSORZIO C.R.E.A.T.E

Website	https://www.create.unina.it/ https://www.create-dih.eu/
Email	Info@create-dih.eu
Phone	+39 081.768.3243
Address	Via Claudio, 21 Napoli, Campania 80125 Italy

ABOUT

As coordinating institution of FP7 and H2020 projects, CREATE is a key player in EU robotics. In addition to wide expertise provided in R&D and technology transfer as a competence centre in I4MS focused on robotics, CREATE offers services including mentoring, technology scouting, feasibility studies, best practice evaluation and dissemination, with a focus on regional SMEs (coordinator: Bruno Siciliano).

- Competence Centres: CREATE, Napoli, Italy (focus on robotics). Additional competence centres are expected in Campania, Lazio, Basilicata and Calabria regions, considering past experience of the CREATE Consortium in the coordination of robotics research projects and technology transfer actions.
- Activities/services: Technical support; Access to users; Training and education; Network development; Ecosystem building and networking; Dissemination and awareness.
- Focus on innovation: TRL 5 – technology validated in relevant environment; TRL 6 - technology demonstrated in relevant environment; TRL 7 - system prototype demonstration in operational environment.

Figure 26: CREATE robot inspecting a pipeline

CURRENT OTHER ACTIVITIES OF CONSORTIUM CREATE

- Computational electromagnetism and applications
- Non-destructive evaluation
- Nuclear fusion technology
- Robotics
- Superconducting magnet technologies
- Conferences & workshops

Figure 27: CREATE robot testing

DIH INFORMATION

Consorzio C.R.E.A.T.E. (CREATE) is a non-profit research organization possessing a legal identity; it belongs, according to the Italian law, to the class "Consorzi", where several legal entities give life to an independent body intended to reach commonly agreed objectives. C.R.E.A.T.E. was founded in late 1992 by ABB-Ricerca and the University of Reggio Calabria with the aim of establishing a stable link between industry and university. The partnership was extended by, successively: Università di Napoli Federico II, Università di Cassino, Università della Campania Luigi Vanvitelli, Università Parthenope, Università della Basilicata and Ansaldo-Energia. The financial resources come from the funds given by the partners at the constitution date and from additional resources gained directly by the partners during their activities.

The scientific and technical aims are pursued by means of the coordinated effort of experts acting as consultants. In addition, the Consortium takes advantage of temporary staff for technical and secretarial support. A collaboration with PRISMA Lab (Projects of Industrial and Service Robotics Mechatronics and Automation) has been active for 15 years. The PRISMA team is committed to pursuing research in robotics and automation and is internationally recognized in the community for their achievements with industrial and service robotics.

Over the years, CREATE has gained vast experience from participation and coordination of twenty FP7 and H2020 European projects in the field or robotics and automation in general and in I&M in particular.

SERVICES

SERVICES RESUME for C.R.E.A.T.E
Business Services: Incubator/accelerator support; Offering housing; Project development; Skills and education.
Ecosystem Services: Community building; Ecosystem learning; Strategy development.
Technology Services: Contract research; Provision of technology; infrastructure; Strategic RDI.

CYPRUS UNIVERSITY OF TECHNOLOGY (CUT)

CYPRUS UNIVERSITY OF TECHNOLOGY

Website	https://www.rcdslab.org
Email	savvas.loizou@cut.ac.cy
Phone	+357 25.00.21.65
Address	45 Kitiou Kyprianou Street, Dorothea Bldg., Office 507, Limassol 3041, Cyprus

ABOUT

The Robotics, Control, and Decision Systems (RCDS) Laboratory and Digital Innovation Hub is part of the Mechanical and Materials Science and Engineering Department of the Cyprus University of Technology (CUT), with a strong focus on education and research. It has extensive collaborations with both industry and RTOs within and outside Cyprus with a record of national, EU and private research funding. The RCDS is a one-stop-shop for all things related to developing and using robotics to support business operations.

Both theoretical and applied robotics are combined to design and build operational systems, coupled with business expertise. CUT is active in the air, ground, underground, marine, and maritime robotics areas, serving multiple business sectors. We work with entities of all sizes (start-ups to multinationals) and academia, government agencies, and NGOs in the EU and beyond.

DIH INFORMATION

The Cyprus University of Technology (CUT) is the legal entity that represents the CUT Robotics Control and Decision Systems laboratory and Competence Centre (CUT-RCDS-CC) DIH. The University was established by law in December 2003, and enrolled its first students in September 2007. It is a new, public, and independent University with high values concerning its academic character, legal status, organizational and administrative structure; it is characterized by its high-level procedure for staff selection, and its relationship with the state. The University aspires to develop itself into a modern, pioneering University able to offer education and high-level research in leading branches of science and technology which have high impact on the economic, technical, and scientific sectors.

RCDS was the first robotics laboratory in Cyprus (established in 2011—we were recognized as a DIH by the EU's Horizon 2020 Framework Programme in 2017). Since then, we have developed externally-funded activities worth more than 26 million euros. We engage in research, technology development, innovation and commercialization activities.

The RCDS Laboratory is at the heart of the RCDS DIH. It provides the foundation (people, expertise, infrastructure) upon which all the services of the DIH are built. The laboratory operates out of two facilities (one focusing on marine and maritime robotics and another one dealing with mobile, aerial and underground robotics) at the Cyprus University of Technology campus.

SERVICES

The research focus of CUT-RCDS-CC is on multi-robot motion task planning, coordination, and control, with applications to mobile, aerial, underwater and underground robots. RCDS-CUT-CC is a one-stop shop for developing robotic systems for real world applications. Its activities cover areas of both academia and industry:

- Education
- Basic research
- Technology development
- Product development
- Innovation support

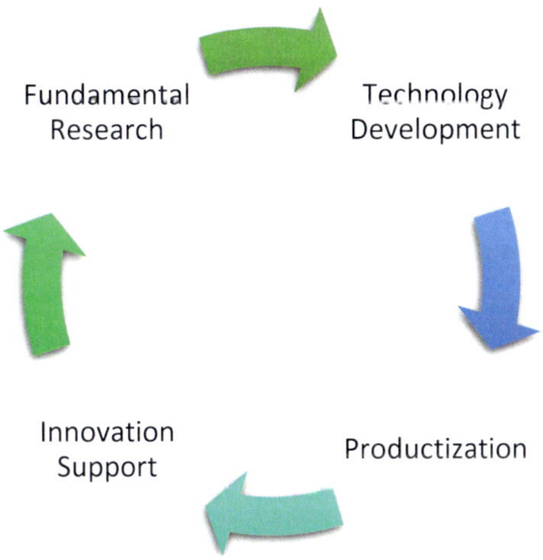

Figure 28: Activities of RCDS-CUT-CC

Our service offering is wide and varied. Are you a startup or a larger corporate entity? Or maybe a government agency, an academic or research entity, or an NGO. Reach out to us to discuss the issues you are facing. Even if we cannot help you ourselves, we might be able to direct you to those who could. We have an extensive, international network of partners capable of addressing a wide range of robotics and automation-related needs

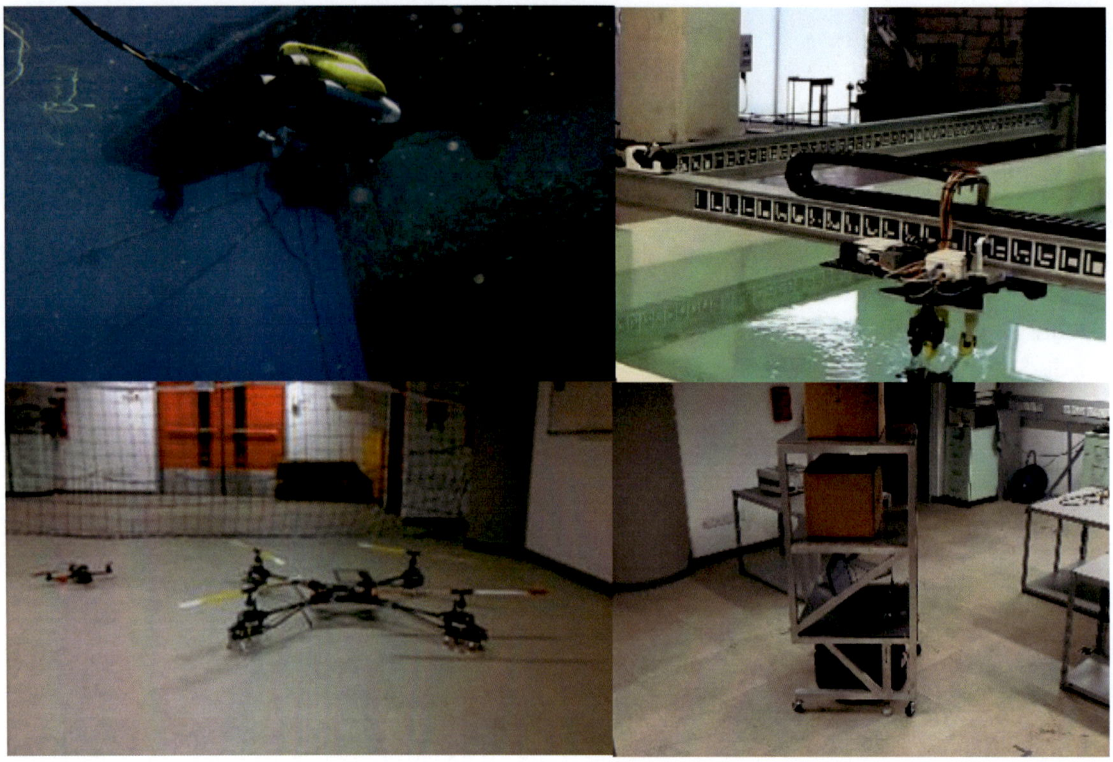

Figure 29: CUT robots for inspection and maintenance

SERVICES RESUME for CUT
Technology, business, and ecosystem services: 1. Strategic RDI. 2. Contract research, technology, and product development. 3. Technical support. 4. Project development support. 5. Incubator/accelerator support. 6. Market analysis. 7. Strategy development. 8. Ecosystem learning. 9. Training and education. **Technological expertise:** Fundamental science and technology development: Decision and control of robotics and autonomous (multi-) agent systems: 1. Navigation. 2. Control. 3. Coordination. 4. Motion task planning. **Applied science and technology development:** 1. Aerial Robotics. 2. Marine and maritime robotics. 3. Mobile robotics. 4. Underground robotics.

THE DANISH ROBOTICS DIH: DTI, OR & DIRA

DANISH TECHNOLOGICAL INSTITUTE

Website	https://www.dti.dk
Email	thkr@dti.dk (Thomas Kroman, Project Manager)
Phone	+45 7.220.2000
Address	Taastrup Gregersensvej 1 2630 Taastrup Denmark

ABOUT

The Danish Technological Institute (DTI) is a self-owned and not-for-profit institution that dates back to 1906. DTI is among the strongest partners when it comes to developing and implementing high-tech solutions for and in (Danish) companies and is a leading research and technology company.

Figure 30: DTI Network

DTI occupies a crucial position at the point where research, business, and the community converge. Its mission is to promote growth by providing innovation, often by adapting good or novel technologies from one place to fit the needs of another setting. DTI is a key partner in ensuring that the green transition takes place in companies in terms of energy, materials, and foods. It has over 1,000 employees, making it one of the world's most significant private institutes to supply approved technological services such as consultancy, tests, certification and training for companies and public-sector organizations. It is organised into over 40 centres of expertise.

DTI's Robot Technology (DTI-RT) centre is located in Odense, the hub of Danish robotics. DTI-RT provides robot and vision solutions for a wide range of areas such as inspection and maintenance, logistics, waste sorting, manufacturing, agriculture, and healthcare. We keep ourselves knowledgeable about the latest tech and about current commercial needs, and we adapt and integrate the tech to meet those needs – either on request or strategically due to our spotting a meetable need without a current solution. Currently DTI-RT also hosts four start-up companies. Our focus is on innovation - applying new research results and robot technologies to create innovative robot solutions for the markets of tomorrow.

Figure 31: DTI facilities

DTI is actively involved in many robotics projects (currently just over 100) in primarily manufacturing, service robotics, and agriculture. In the context of this project, the most relevant is probably our suite of quality assurance technologies and solutions. DTI specializes in the more complex flexible solutions that require feedback from sensors, especially cameras and force sensors. DTI is state of the art in sensor-based control because we can make use of methods that have not yet entered the commercial libraries. We provide our customers with individual consultancy, industrial solutions, courses for both blue- and white-collar workers covering various aspects of robotics and automation, and networking opportunities, as well as performing applied research and development. Our continued focus is on creating positive impact for the end users of robotic technologies and being first mover on developing and deploying new breakthrough solutions on the market and in as many branches and domains as possible.

DIH INFORMATION

The Danish Robotics DIH is based in DTI Robot Technology and comprises mainly DTI, the Danish Robot Association (DIRA), and Odense Robotics (OR). The members of DIRA and OR include nearly all the robotics developers in Denmark: industrial, academic, and integrators. We help large companies optimize their production, but we also house emerging entrepreneurs and provide robot facilities and expertise for both robot developers and for end users. DIRA and DTI together run an annual robotics exhibition ROBOTBRAG that showcases the best new and upcoming robot solutions in Denmark.

DTI-RT employs specialists in cobot safety, handling and process robotics, quality control and inspection, service and mobile robotics and production optimization. Our three robot innovation halls are living labs, which work both as an inspiring visitor center and as a focal point for the Institute's technical development and testing of innovative technology for both industry, production, welfare, and service.

Odense Robotics is Denmark's national robot, automation and drone cluster. Our vision is to make Denmark an even bigger, better robot nationis by accelerating innovation and sustainable development in industry. We offer companies opportunities to develop innovative products and technologies, and strengthen their foundation for growth through projects, networks and events. Odense Robotics is a national cluster organisation with headquarters in Odense and regional hubs across the country.

DIRA, the Dansk Robot Network, was founded in 1982. DIRA today has more than 215 member companies. DIRA is an active network for developers, suppliers, buyers, end-user organisations, educational institutions and authorities with an interest in robotics. As a member of DIRA, you get a unique framework that you can use as a platform to create strong network relationships and expand your knowledge of the robotics and automation industry.

SERVICES

SERVICES RESUME for the Danish Robotics DIH
Business services: Consultancy; Incubator/accelerator support; Access to finance; Individual roadmaps; Skills and education.
Ecosystem Services: Community building; Ecosystem learning; Strategy development.
Technology Services: Development of tailored solutions; Vision-based quality control; Cobot solutions; Contract innovation; Strategic RDI; Testing and validation.

DIH PARIS-SACLAY: DIGIHALL, CEA & SYSTEMATIC

FRENCH ALTERNATIVE ENERGIES AND ATOMIC ENERGY COMMISSION

Website	https://www.cea.fr/
Email	N/A
Phone	N/A
Address	France

ABOUT

CEA is a government R&D agency whose field of expertise applies to energy, health, transport, security, defence, and ICT. CEA has more than 20,000 researchers working throughout France.

The Robotics Unit inside CEA LIST provides the technological expertise in robotics for inspection and maintenance. The Robotics Unit was founded over 40 years ago. Specialized originally in remote handling for operations in hazardous environments (nuclear and underwater applications), the Unit now focuses on manufacturing applications. It also holds an historical background in robotics for healthcare applications including assistive, surgical robotics and rehabilitation robotics. Its main research theme is human-robot collaboration (co-working). The technological research covers mechatronics and the conception of actuators for the design of innovative robotics systems, robot control, supervision and user assistance (force, haptic, vision, graphic, immersive feedback) aiming at an efficient human-robot collaboration in all application domains. CEA is deeply involved in innovation and development of the European industry. CEA is the most innovative public research organization in Europe and 2nd worldwide according to the CLARIVATE ranking[2].

The Unit has regular collaborations with industrialists in France and abroad, supporting large groups or SMEs. CEA is involved in the creation of spin-offs, such as Haption [2] (Haptics devices), iSYBOT [3] (collaborative robotics), DIOTA [4] (solutions for manufacturing), Light and Shadow [5] (virtual reality).

CEA has a strong culture for defending Intellectual Property Rights. CEA holds a large number and panel of patents related to robotics and ICT. CEA's policy is to transfer licences on the patents it holds to industry. The objectives of CEA regarding IP are to protect knowledge and offer a context for building trust with industry using CEA's licences.

2 Clarivate top 100 innovators, https://clarivate.com/top-100-innovators/the-top-100/cea-french-alternative-energies-and-atomic-energy-commission/

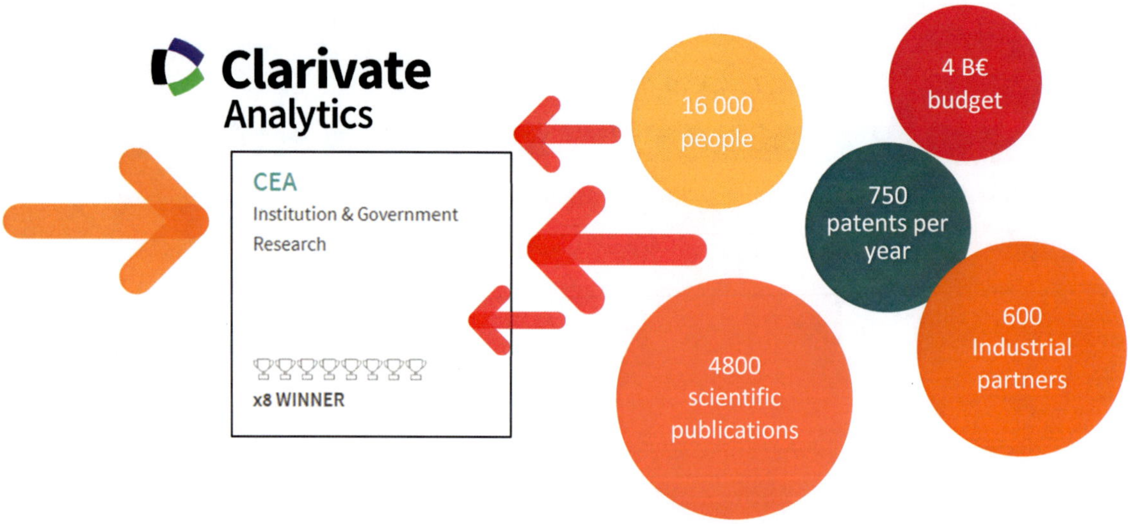

Figure 32: Innovation and research at the CEA in a few figures

The Robotics Unit is composed of 50 people, half being permanent. Most of the personnel are experts in robotics—full time researchers in robotics from PhD students to senior researchers and project managers. The Unit also holds senior business developers who develop the technical and commercial activity and the bilateral links with industrialists.

DIH INFORMATION

CEA is part of DIGIHALL, the Paris-Saclay DIH for robotics and inspection and maintenance. The DIH description comes after Systematic's entry (described below).

SERVICES RESUME for CEA
Business Services: Incubator/accelerator support; Offering housing; Project development; Skills and education. **Technology Services:** Contract research; Provision of technology; Infrastructure; Strategic RDI; Technical support; Testing and validation. The DIH offers a broad set of technologies resulting from its research activities. **Ecosystem Services:** Community building; Ecosystem learning; Strategy development.

DIH PARIS-SACLAY: DIGIHALL, CEA & SYSTEMATIC

SYSTEMATIC PARIS REGION

Website	https://systematic-paris-region.org/
Email	contact@systematic-paris-region.org
Phone	+33 16.908.0683
Address	Site Nano Innov Palaiseau IDF 91120 France

ABOUT

Systematic Paris-Region Deep Tech cluster brings together and runs an ecosystem of excellence of 900 members dedicated to Deep Tech (Cyber & Security; Data Science & AI; Advanced Engineering & Computing; Digital Infrastructure; Open Source; Optics & Photonics) and to major economic and societal challenges.

Since its creation in 2005, Systematic has brought together and led a community of nearly 900 members, including 600 start-ups and SMEs, 140 major groups, 140 academics, a college of investors and a college of 20 local authorities.

Systematic connects stakeholders in the software, digital and deep tech industries through collaborative innovation, networking and technology sourcing, aiming to:

- Create and design new products, services and uses, using an open innovation approach;
- Bring member innovations to market;
- Accelerate the growth and competitiveness of companies;
- Support the economic development of the region.

In terms of its core business, supporting the creation of collaborative or single-partner R&D projects at both regional, national and European levels, Systematic is France's leading cluster, with 626 projects supported through to funding, representing a total of 3.18 billion euros in R&D costs, and 620 products and services resulting from this R&D.

Systematic Paris-Region is supported by the French government (DGE, Ministry for Armed Forces, Defence Innovation Agency), the Paris Region and local authorities.

Systematic's operational team is 40 FTE in staff size and covers different skill sets: communication, innovation, project management including EU-funded projects, and funding mechanisms.

RIMA gave Systematic the opportunity to connect its ecosystem to other Deep Tech ecosystems in Europe. Mentoring robotic experiments funded through RIMA's open calls, especially drone-related experiments, stressed the importance and relevance of Systematic's Hub Drone flight test facilities.

Systematic is a founding member of DIGIHALL, the Paris-Saclay DIH.

System know-how in clustering Deep Tech startups, companies, research centres, groups and public institutions. Know-how in dissemination and communication actions in regional, national and European projects.

Capacity building and new capabilities concerning European project coordination, topic proposal identification and implementation.

Organization of drone experiments and demonstrations. Systematic's Hub Drones manages the largest air space dedicated to drone operators in the Paris Region.

DIH INFORMATION

DIH Paris-Saclay for robotics and I&M proposes a set of services to favour digitizing of industry and develop innovation in different sectors of I&M for infrastructures. All members have offices on the same site (Nano Innov). DIH Paris-Saclay offers a set of technologies and expertise in robotics and related technologies that are not available on the market. DIH Paris-Saclay has connections with the major research organizations in robotics, AI, and big data in France. It has developed many links with techno and service providers in robotics. It maintains regular contacts with industries, large groups, asset owners and operators involved in France and in Europe in various sectors, including the nuclear industry, water supply and sanitation, energy distribution and production, and transport (rail, routes, waterways, maritime, aerospace.

The geographical scope is Paris region first. It extends to the national territory and Europe and also has the capacity to address the international market.

SERVICES

DIH Paris-Saclay offers a set of technologies and expertise in robotics, AI, non-destructive testing, additive manufacturing, communication, embedded architectures, software engineering and related technologies like computer vision, scene understanding, natural language processing. The DIH has also a strong expertise in human factors and human machine interactions.

The services proposed by the DIH aim to facilitate deployment and uptake of technologies, building connections and trust and understanding the needs of industrialists in order to promote economic development.

SERVICES

SERVICES RESUME for DIH Paris-Saclay

Business Services:

- *Access to finance*: Facilitate connections between investors (private investors, investment funds, business angels or corporate) and SMEs and startups. Meetings with experts in funding of research, development and innovation.
- *Project development*: Guide customers through funding schemes and related calls for projects and enable them to be more efficient by targeting only suitable calls. Systematic offers a selection of calls for projects. Throughout the year, customers are kept informed of calls likely to be of interest to them, directly linked to the themes and issues specific to our ecosystem, at regional, national and European levels. Calls for proposals can be found on Systematic's website, but also in newsletters and are presented in monthly webinars.

Technology Services:

- *Provision of technological infrastructure*: Systematic's Hub Drone indoor and outdoor flight test facilities "Centre d'excellence drones Ile de France (CEDIF)". CEDIF is supported by three main stakeholders: Thales, Eurocontrol and Systematic, with the aim of offering a complete platform to incubate, validate and industrialise all direct and indirect drone activities. CEDIF is located between the towns of Saint-Quentin en Yvelines (site of the 2024 Olympic Games) and Brétigny sur Orge (re-use of a former airbase). CEDIF offers technical support on site regarding access to and use of the flight test facilities (indoor and outdoor).

Ecosystem Services:

- *Brokerage, Matchmaking*: Organization of meetings between stakeholders in the value chain to bring demand and offer sides together to share ideas and analyse opportunities.
- *Ecosystem building*: Ecosystem building and dynamization of networking is at the heart of Systematic's mission. Systematic's ecosystem is centered around Deep Tech in the Paris-Region and brings together start-ups, SMEs, scale-ups, large groups, research, and higher education institutions.

GERMAN AEROSPACE CENTER (DLR)

GERMAN AEROSPACE CENTER (DLR)

Website	https://www.dlr.de
Email	rima@dlr.de
Phone	+49 (0) 151 41 934 924
Address	20 Münchener Straße Weßling BY 82234 Germany

ABOUT

The German Space Agency undertake statutory tasks in the space sector on behalf of the German Federal Government. Under the "Raumfahrtaufgabenübertragungsgesetz" and within the scope of the tasks effectively assigned to us, we implement the space strategy of the Federal Government, develop and manage the national space programme, and represent the interests of the Federal Republic of Germany in space-related international bodies in accordance with the tasks assigned to us. We advise the Federal Government and develop initiatives and strategic approaches for space policy. All our activities are based on the guidelines laid down by the Federal Government in its space strategy.

DIH INFORMATION

The DIH is based on the Institute of Robotics and Mechatronics at German Aerospace Centre (DLR) located close to Munich. The Institute of Robotics and Mechatronics develops a wide array of robots to enable humans to interact more safely and efficiently with their surrounding environments.

Figure 33: DLR building

The robots are designed to act in surroundings inaccessible or dangerous to humans as well as to support humans in everyday life and work. Around 240 full time researchers at the institute have been working on different robotics projects using a wide network of partner companies in Germany and in Europe. Robotics for inspection and maintenance is an important part of field robotics, one of the key research areas of the institute. DLR Institute of Robotics and Mechatronics is embedded in different local networks for industrial research and development. The institute is involved in DLR wide technology transfer activities and supports the establishment of start-ups. The DIH coordinates the activities related to industrial inspection and maintenance.

The main building of the institute contains the main labs with robotic systems and rooms for meetings and presentations. Several testing areas for field robotics experiments are available on campus. The campus is next to a local airport, where different experiments with aerial systems can be conducted.

SERVICES

SERVICES RESUME for DLR
Business Services: Incubator/accelerator support; Project development. **Technology Services**: Contract research; Provision of technological infrastructure; Strategic RDI. **Ecosystem Services:** Community building; Strategy development

GERMAN RESEARCH CENTER FOR ARTIFICIAL INTELLIGENCE (DFKI)

GERMAN RESEARCH CENTER FOR ARTIFICIAL INTELLIGENCE

Website	https://robotik.dfki-bremen.de/en/startpage.html
Email	robotik@dfki.de
Phone	+49 421 17845 4100
Address	Robert-Hooke Str. 1, 28359 Bremen.

ABOUT

The German Research Center for Artificial Intelligence (DFKI) was founded in 1988 as a non-profit public-private partnership. It has research facilities and offices throughout Germany, in Kaiserslautern (headquarters), Saarbrücken, Bremen, Berlin, Osnabrück, Oldenburg, Darmstadt, Lübeck, and Trier. In the field of innovative commercial software technology using Artificial Intelligence, DFKI is the leading research center in Germany.

DIH INFORMATION

The DIH in Bremen is currently centered around the DFKI Robotics Innovation Lab (RIC), which is one of the leading RTOs for inspection & maintenance robotics in the region and in Germany. DFKI RIC comprises close to 100 full time researchers, most of them with extensive experience in robotics R&D for and with customers from industry and SMEs. Most technical experts at DFKI also have experience with teaching and academic education. DFKI RIC maintains a team dedicated to project development and project management and has access to technical, administrative, and legal experts in other DFKI institutes throughout Germany.

Figure 34: DFKI building

DFKI is embedded in a local innovation ecosystem that is currently evolving dynamically, building up services for innovation management, start-up support, vocational training, application-oriented education etc.. In addition, DFKI is developing capabilities for the commercial exploitation of research results and the commercial offering of training courses. Both these internal capabilities and the offerings through the innovation ecosystem will be leveraged by the Bremen DIH in the future.

Figure 35: DFKI facilities

The main DFKI building in Bremen hosts office space and meeting rooms for the DFKI employees, but potentially also for robotics start-ups. The building was significantly extended in 2021 and now features a wide array of labs and testing facilities. This includes a pressure test lab, a virtual reality lab, and the unique large-scale maritime exploration facility (see below). Labs for other application domains, such as space and extreme terrestrial environments, are included as well.

The Bremen DIH for robotics for inspection and maintenance offers a set of services to support the digitization of industry and the development of innovation in different sectors of infrastructure inspection and maintenance. DFKI can provide technologies and expertise in robotics and related technologies (e.g., AI), as well as technical infrastructures that are otherwise not available on the market. The services offered by the DIH are mostly technology services, but some ecosystem and business services are also provided.

SERVICES

SERVICES RESUME for DFKI
Business Services: Incubator/accelerator support; Project development; Skills and education. **Ecosystem Services:** Community building; Ecosystem learning; Strategy development. **Technology Services**: Contract research; Provision of technology; infrastructure; Strategic RDI; Technical support; Testing and validation.

PIAP HUB: ŁUKASIEWICZ -PIAP

ŁUKASIEWICZ RESEARCH NETWORK - INDUSTRIAL RESEARCH INSTITUTE FOR AUTOMATION AND MEASUREMENTS PIAP

Website	https://piap.lukasiewicz.gov.pl/
Email	krzysztof.lipiec@piap.lukasiewicz.gov.pl jakub.glowka@piap.lukasiweicz.gov.pl
Phone	+48 22.874.0164
Address	Jerozolimskie 202 Warsaw 02-486 Poland

ABOUT

Łukasiewicz -PIAP (ŁPIAP) is a research institute established in 1965 with strong market presence in the area of robotics. Currently it has around 250 employees, of which more than 2/3 are scientists and/or engineers. ŁPIAP is a part of Łukasiewicz Research Network, which has more than 7000 employees in 12 cities in Poland. The main objective of the Łukasiewicz Network is to conduct research work of key importance to the national policy and commercialize the results. Within the Łukasiewicz Network, ŁPIAP specialises in Digital Transformation and Intelligent Mobility. Our leading active topic is robotics in the application domains: defense and security, inspection and maintenance, and manufacturing.

Almost 60 years of close cooperation with industry have resulted in a number of new developments and significant implementations. ŁPIAP has been providing services to domestic and foreign companies under the brand PIAP HUB (no. 1408 in the S3 portal) since 2017. During this time, it participated in 4 H2020 consortia creating DIH networks, developing service standards, and implementing support programmes. ŁPIAP has also participated in the activities of other DIH networks, is one of the founders of the European DIH network in the area of artificial intelligence and participated in a mentoring programme for DIH managers organised on behalf of the European Commission by PWC and Oxentia. In 2020, the Institute became the EIT Manufacturing Hub in Poland.

The original partnership of PIAP HUB has grown and merged with other DIHs and major research and industrial organisations from the Mazovian region to establish DIH Mazovia, which is now an EDIH.

DIH INFORMATION

For over 25 years ŁPIAP has been active in research collaborations within the European Union, participating in every framework programme since FP4, and including EDF, ESA, EDA, NATO and Digital Europe programmes.

This activity was extended via a growing number of research results being implemented in SMEs and large enterprises alike, mostly in the manufacturing sector. ŁPIAP has spun out PIAP–Space, one of the leaders of the Polish space sector. Last but not least, in the Defence and Security domain ŁPIAP is one of the biggest European producers of IEDD/EOD/CBRN mobile robots and accessories. These robots, developed in-house, are on special forces equipment in 25 countries around the world.

Being a part of RIMA gave us easy access to the highest-performing R&D institutions in Europe, which translated into benefits for our most demanding customers, who required a scale of research activity or kinds of expertise and/or infrastructure not available locally. On the supply side, ŁPIAP gained valuable additional experience in international R&D services for SMEs through our involvement in RIMA. The format of services has been adjusted to take account of the good practices of the network partners, which helps a lot when wider cooperation is required. The focus on I&M in communicating the RIMA project resulted in the accumulated attention of relevant industrial partners in Poland, namely infrastructure operators and potential solution providers that could give an additional spin to the uptake of robotics in this field.

Przemysłowy Instytut Automatyki i Pomiarów PIAP (ŁPIAP) became a member of the Łukasiewicz Research Network - the third largest research network in Europe, bringing together 32 Polish institutes - in 2019. The main objective of the Network's activities is to conduct research work of key importance to the country's policy and to commercialize its results. ŁPIAP specializes in digital transformation and intelligent mobility. With the establishment of the Network, ŁPIAP has gained access to a number of tools, procedures and experts in the field of organization, support for enterprises, as well as an incomparably greater possibility of informational and promotional impact.

ŁPIAP's annual revenues reach PLN 60 million, including:

- Funds obtained from commercial activities on the Polish and foreign markets 73%,
- National funds for research and development 7%,
- National funding instruments 11%,
- International funds for research 8%.

Since 1993, ŁPIAP has been involved in over 100 international projects, including coordinating the FP7 TALOS project worth EUR 20 million. Subsequent projects and closer cooperation with companies has resulted in the Institute adding new services to its offer, as described later in this document.

The Institute has a rich research, design and construction and executive potential. As of 30 September 2020, ŁPIAP employed 278 people, including 201 people in the support division and 77 people in the research division.

ŁPIAP has a team of 8 people whose main task is accounting for grants from European and national funds, including handling inspections and audits, whose employees have between 7 and 20 years of experience in this area. The team is able to carry out financial management of several dozen projects in parallel.

The infrastructure needed to manage a complex project and provide information, promotion and account management services is mainly hardware with office software (including a set of Microsoft cloud services and local infrastructure), high-speed Internet connection, client databases (a joint Łukasiewicz CRM (Customer Relationship Manager) is soon to be launched) and software listed further on in the description of individual services. ŁPIAP is also the publisher of the monthly magazine Automatyka, which is subscribed to by 4000 specialists in the field of automation and industry.

Its facilities include a conference centre and rooms for smaller group meetings, demonstration spaces in various configurations, indoors and outdoors, as well as a car park for visitors, located in a large office and technology complex at Aleje Jerozolimskie in Warsaw with a total area of 20,000 m^2 on a 7 ha site.

ŁPIAP has extensive research and technical facilities which are used for services of a technical nature required during the execution of research projects. These include numerous specialized laboratories, many of which are unique in the country and equipped with the latest generation of equipment. The laboratories are designed to carry out specialized research, mainly in the field of robotics, as well as research of a universal nature that takes place in many research projects. An important element of the technical facilities is the in-house test facility, which enables the making of models and prototypes of the devices under investigation and their production at small scales.

SERVICES

The experience in research, innovation and commercial high-tech activities evolved into a Digital Innovation Hub, which was launched in 2018 grew further into an EDIH in 2023. It aggregates and makes available the most relevant services of ŁPIAP and its partners and leverages them with public funding.

Depending on customer's requirements, feasibility study, proof of concept, demonstration, and technology implementation (both pilot and full-scale) services are available in the areas of product development, manufacturing technologies or system integration, directly from the ŁPIAP labs. A much larger portfolio of services is available from the DIH network in a one-stop shop format.

The technology services are backed up with business, funding and networking offers as well as thematic events and technology-related training of various kinds.

SERVICES RESUME for ŁPIAP
Business Services: Access to finance; Incubator/accelerator support; Skills and education. **Ecosystem Services:** Community building; Ecosystem learning; Strategy development. **Technology Services:** Contract research; Provision of technology; Infrastructure; Technical support.

LABORATORY FOR MANUFACTURING (LMS)

LABORATORY FOR MANUFACTURING (LMS)

Website	http://lms.mech.upatras.gr
Email	makris@lms.mech.upatras.gr karagiannis@lms.mech.upatras.gr
Phone	+30 261.091.0160
Address	University of Patras Rio, Patras 26504, Greece

ABOUT

The Laboratory for Manufacturing Systems & Automation (LMS) focuses on research and development in cutting-edge scientific and technological fields. LMS is involved in a number of research projects funded by the CEU and European industrial partners. Particular emphasis is given to co-operation with European industry in general and especially "hi-tech" firms. It currently employs more than 100 researchers organized in four groups: a) Manufacturing Processes Modelling and Energy Efficiency, b) Robots, Automation and Virtual Reality in Manufacturing, c) Manufacturing Systems d) Software Development for Manufacturing.

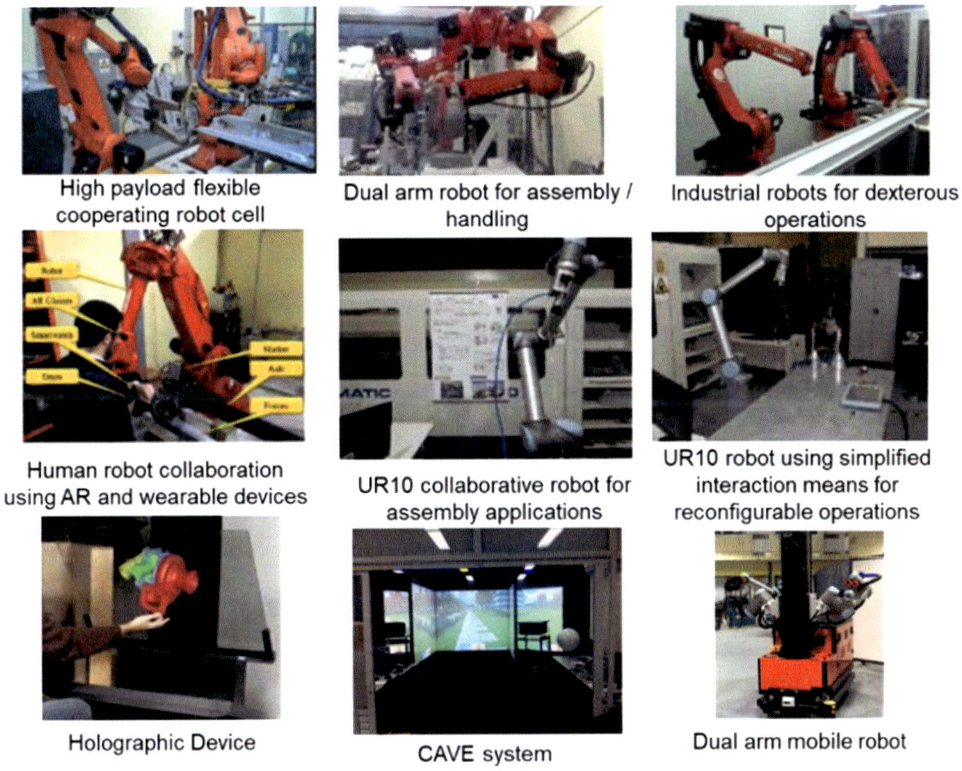

Figure 36: Equipment in LMS facilities

In terms of LMS as a DIH, it is actively engaging SMEs in robotic technologies in the manufacturing sector at both regional and national level. Using its long experience with HRC technologies and its activities on the European landscape, LMS transfers its knowledge to Greek SMEs and to large industries, closing the gap between the Greek and the European manufacturing competences.

DIH INFORMATION

LMS runs its robotic group activities in a fully-equipped machine shop, where the following robotics cells are installed and used for research:

- Flexible cell with two heavy duty industrial robots (COMAU NJ130 and NJ370) equipped with fixtures, grippers, welding gun, RFID sensors and other equipment.
- Dual arm robot cell equipped with PILZ SafetyEye, human-robot interaction sensors, Schunk grippers.
- Cooperating robots' cell for small payload assembly operations equipped with two COMAU RACER 7-1.4, conveyor, grippers and camera system.
- Human-Robot cooperative cell:
  - This cell enables cooperation between a human and a high-payload robot (COMAU NJ130).
  - The cell is equipped with a state-of-the-art 3D scanner (SafetyEye) and the latest robot safe move functionality.
  - The cell provides the operator with wearable devices and infrastructure promoting safe and intuitive interaction between the human and the robot.
- UR10 robot cell for Human Robot Collaboration equipped with grippers for pick & place tasks.

LMS also has experience with VR and other HRI technologies, possessing a wall CAVE system BARCO RLM-W12 projectors and 4x VICON Bonita 3 tracking cameras, several Kinect sensors, industrial voice-canceling headsets and leap motion sensors.

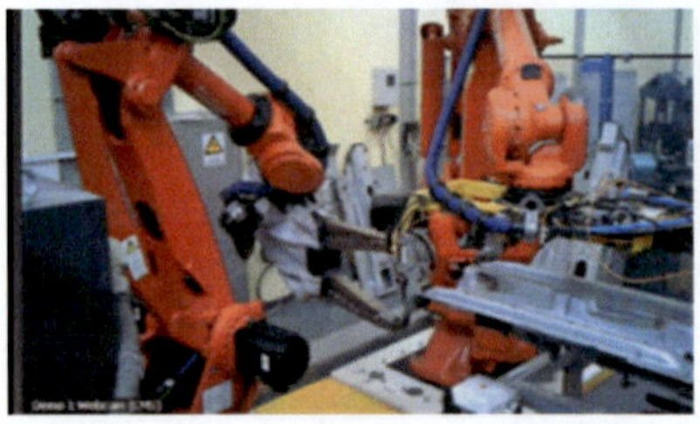

Figure 37: High payload flexible cooperating robots cell

Figure 38: Dual arm robot for assembly/handling

Figure 39: Industrial robots for dexterous operations

Figure 40: Human-robot collaboration using AR and wearable devices

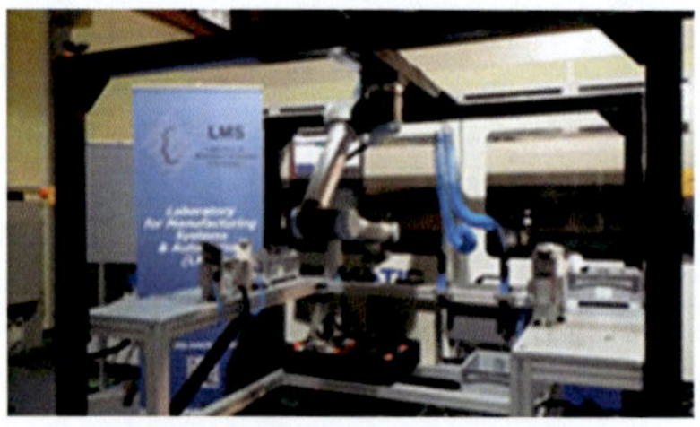

Figure 41: UR10 collaborative robot for assembly applications

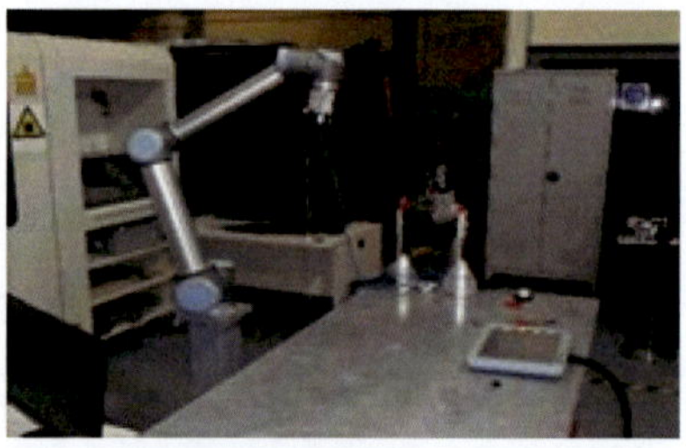

Figure 42: UR10 robot using simplified interaction means for reconfigurable operations

Software tools used in industry are available in LMS, such as product development (Pro/Engineer, CATIA), for process planning and simulation (DELMIA, Process Simulate, IMPACT, WITNESS. ROBCAD, Invision, WinMod), for integration/communication and visualization Virtools and a Virtual and Augmented Reality Room.

LMS IN RIMA

LMS running RIMA services reinforces European innovation by helping SMEs and slightly bigger companies both financially and through consulting services to bring their solutions closer to the market. The SMEs received mentoring support through a variety of business, technical and ecosystem services, as described in the section above. The long expertise of the RIMA DIHs and specifically of LMS who had a leading role organizing the mentoring activities, helped the companies to improve their products and services, create new ones, receive extra funds and increase their employee numbers. In some cases, these other kinds of support had even higher value for the SMEs than the funding itself.

From LMS side, RIMA provided a great opportunity to familiarize ourselves more with the struggles SMEs often face during their development, both on technical and business sides. Examples of such challenges are the lack of resources, the obstacles in networking, the absence of contacts with large industries, the lack of knowledge when it comes to certification, standardization, technology scale-up, etc. Overall, SME acceptance of the RIMA programme has been very positive, leading to the completion of all applications despite unexpected problems such as the pandemic causing delays, without any stoppage from anyone. Beyond the funding, the RIMA DIHs covered the need for advice and facilities, which was greatly appreciated by the SMEs and has been identified as key benefit for them.

More specifically, among the different services offered by the RIMA network, the following are the ones most frequently requested from LMS:

- Technical and business advice.
- Access to infrastructure.
- Support on standardization, IPR, legislation etc.
- Dissemination and networking opportunities.

The knowledge and expertise of LMS and the RIMA DIHs in general was recognized by the applicants, so they tried to make the most of the opportunity and requested services beyond the level of guidance and mentoring. Many applicants benefitted from accessing DIH facilities, gaining hands-on experiences and testing if their products conform to market expectations. This was made easier by the mentoring program being designed and organized by LMS in a flexible way. LMS understands that each SME faces different restrictions, such as level of knowledge and experience, legislation restrictions, financial capabilities etc., so each DIH should adapt and try to mentor them accordingly. Part of the success of this organization was due to the high availability of partners to respond to mentoring requests and the highly supportive and tolerant spirit towards the SMEs. The mentors were advised to be flexible both from an operational and an organizational point of view, to refrain from being judgemental towards the SMEs, to try to have an open mind to their needs. The findings from real practice showed that the DIHs played an important role for the SMEs. Success stories shared in RIMA events show the benefits for SMEs participating in the RIMA scheme, such as the acquisition of new knowledge both on business and technical level, the support offered in testing new technologies, and the networking capabilities offered on a pan-European level.

SERVICES

SERVICES RESUME for LMS
Business Services: Ecosystem learning – Community building; Strategy development. **Ecosystem Services:** Access to finance – Incubator/accelerator support; Project development; Skills and education; Offering housing. **Technology Services**: Contract research; Provision of technology & infrastructure; Technical support; Testing and validation.

SINTEF & RINVE

SINTEF

Website	www.sintef.no
Email	Aksel.A.Transeth@sintef.no Mariann.Merz@sintef.no
Phone	+47 40005100
Address	SINTEF, P.O. Box 4760 Torgarden, NO-7465 Trondheim, Norway

ABOUT

For over 70 years, SINTEF3 has developed solutions and innovation for society and customers all over the world. SINTEF is based in Norway and is one of Europe's largest independent research organizations (about 2200 employees) and covers research and innovation in technology and social science. SINTEF is a not-for-profit organization, i.e., no owners receive any form of dividend. Any financial surplus is invested in scientific equipment, skills, and expertise.

SINTEF is made up of 6 research institutes, each dedicated to one of the following thematic topics: Community, Energy Research, Industry, Manufacturing, Ocean and Digital. In addition, SINTEF Holding manages SINTEF's ownership in start-up companies and other enterprises. SINTEF is

3 https://www.sintef.no/en/sintef-group/this-is-sintef/

primarily located in Trondheim (1400 employees) and Oslo (400 employees). We also operate a research station in Svalbard.

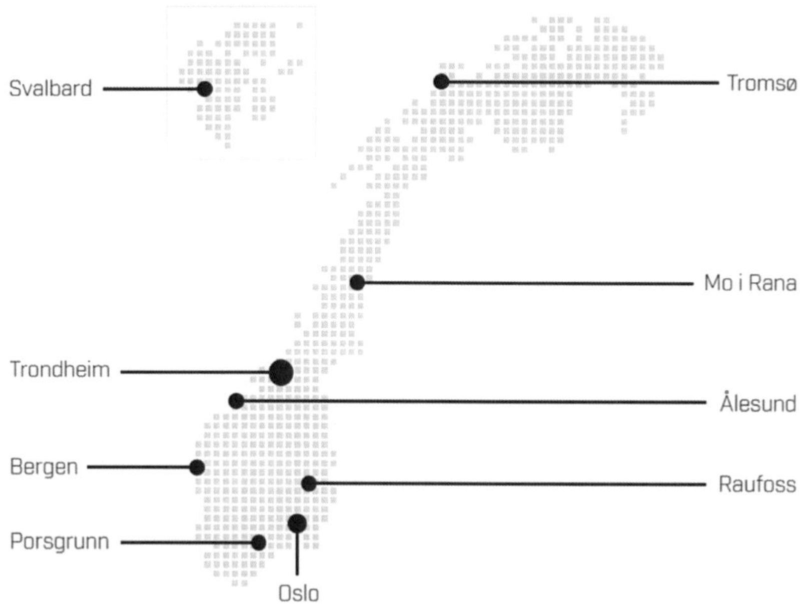

Figure 43: SINTEF network

SINTEF also has personnel in offices and subsidiary companies in Bergen, Tromsø, Ålesund, Raufoss, Mo i Rana, Porsgrunn, Bergen, Brussels and Hirtshals (Denmark). Our management system is certified (by DNV) to be in accordance with ISO 9001:2015, ISO 14001:2015 and ISO 45001:2018.

SINTEF was originally founded by what is now known as the Norwegian University of Science and Technology (NTNU) and has collaborated closely with NTNU since then, including sharing key laboratory infrastructure. SINTEF also has a collaboration agreement with the University in Oslo (UiO) and other prominent universities in Norway and internationally.

Robotics and autonomy are key research areas for SINTEF, including operational environments underwater, on the surface, on the ground, in the air and in space. SINTEF Digital is the main partner from SINTEF in RIMA and leads our research and development within robotics and autonomy. This interdisciplinary field often involves collaboration with domain experts from the other institutes.

In terms of SINTEF as a DIH with activities on inspection and maintenance robotics, a national network on I&M robotics called RINVE is central. RINVE was initiated by SINTEF, co-funded by the Research Council of Norway, and has around 30 members. The member list comprises of end users (asset owners), technology providers, service providers, research institutes and universities.

Figure 44: SINTEF underwater testing facilities

DIH INFORMATION

SINTEF is one of Europe's largest independent research organizations. We generate **new technologies and knowledge** together with our clients. SINTEF is an independent, not-for-profit organization. None of its owners receive any form of dividend. Any financial surplus is invested in scientific equipment, skills and expertise. We work extensively with other research organizations in Europe. SINTEF has an office in Brussels and is by far the largest Norwegian participant in the EU's research programmes.

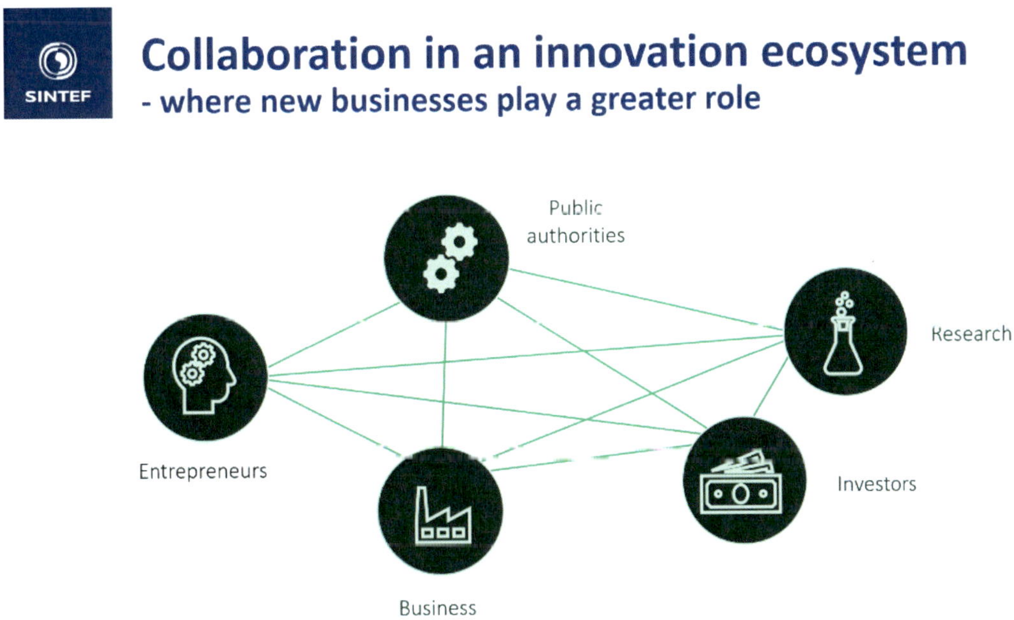

Figure 45: SINTEF as part of an innovation ecosystem

We offer world-leading **laboratory and test facilities** in a wide range of technological fields. Our advanced laboratories range from microelectronics and nanoscale technologies to the world's largest laboratory for the multiphase transport of oil and gas, and the world's largest marine laboratory.

The **commercialization of our research results** is part of SINTEF's role in society. We achieve this by means of licensing and creating new companies based on technologies developed as part of our research activities.

Several institutes and research groups at SINTEF contribute to SINTEF's overall service delivery within **robotics for inspection and maintenance.** We have institutes specializing in technology and applications within specific sectors (e.g., SINTEF Manufacturing, SINTEF Ocean, SINTEF Industry, SINTEF Energy) as well as an institute focusing on enabling technologies (SINTEF Digital). There is cooperation between the institutes and research groups to deliver a complete set of technology- and application- specific know-how and technologies. We have over 50 people working on robotics-related technologies including business development.

RINVE[4]

RINVE is a network of Norwegian end users, asset owners, service providers, technology/system providers, research organizations and universities within robotics and automation in inspection and maintenance. RINVE has around 30 members and is growing its member base.

The purpose of RINVE is to facilitate Norwegian organizations' ability to capitalize on the possibilities within I&M robotics. A main service of RINVE is to provide innovation workshops where end users can initiate collaborations with providers and research organizations to solve key problems in their industry.

The member base of RINVE currently mainly focuses within oil and gas (topside and subsea), transportation (rail, road, aviation, harbors), maritime, and energy. The members together have offices throughout Norway.

RINVE is co-financed by its members and was co-financed by the Research Council of Norway until 2021.

SINTEF has identified 12 key areas of expertise that are shown in the following figure:

Figure 46: SINTEF areas of expertise

Robotics for inspection and maintenance is relevant for many of the key areas meaning personnel from domain-specific research groups often work with personnel from SINTEF Digital that has robotics as a core competence. In addition to the domain-specific knowledge, key strengths within the field of robotics include path planning, control algorithms, machine vision, AI, prototype development and optimization.

EXPERTISE AND SCOPE

In 2021, SINTEF carried out 6900 research projects for 3200 clients. Our multidisciplinary approach enables us to create teams with the broad range of expertise our customers and projects often demand. These projects span all our areas of expertise shown above. For the specific research area of robotics and autonomy, projects range from industrial robotics applications to the development of air traffic management systems, and involve robot vision, estimation, localization, navigation, path planning, optimization, system integration, use-case analysis, and analysis of regulatory and

4 https://www.sintef.no/RINVE/

safety implications. Recent and ongoing projects that are relevant for I&M include Seavention, SFI Harvest, ROBPLAN and PILOTING.

In the Seavention project (2018 – 2021), we researched robot autonomy for subsea operations such as I&M. On the topic of planning, the project considered automated planning for unmanned underwater vehicles (UUVs) with human oversight but minimal interference.

Centre for Research-based Innovation (SFI) Harvest (2020 – 2028): Here we develop knowledge and technology to harvest and process low trophic marine resources. As part of the SFI, SINTEF is involved in ongoing work on plan execution for multiple UUVs.

In the ROBPLAN project (2021 – 2024), we research and develop methods for AI planning to enable robust autonomous robots, with an application area for mobile manipulators and UAVs for inspection and maintenance.

Figure 47: SINTEF robot tests

In the H2020 project PILOTING (2020 – 2023), we pursued the adaptation, integration, and demonstration of robotic solutions in an integrated platform. These systems were tested and evaluated in three large-scale pilots (refineries, bridges/viaducts and tunnels). SINTEF specifically worked with data collection, sensor fusing and positioning.

When appropriate, SINTEF commercializes our research results e.g., through licensing the technology and creating new companies. SINTEF has spun out several new companies including Zivid Labs, a company specializing in sensors for producing high-quality 3D data relevant for e.g., inspection, manipulation and quality check operations.

LESSONS LEARNED FROM RIMA

Through the work performed as part of the RIMA network, SINTEF has greatly expanded its network within the area of inspection and maintenance robotics, forming new working relationships with other European DIHs and relevant SMEs. Furthermore, the project has contributed to an increased consciousness and structure regarding the large portfolio of services that an RTO can deliver (and actually delivers), including what a service is, what the different services are, and the networking aspects of collaborating on delivering services.

Additional key takeaways include:

- How a cascading funding structure like FSTP can be a useful innovation tool when properly implemented and managed.
- Inspiration to offer new services based on the successes reported by other DIHs.
- A better feel for what types of technology SMEs are working to develop and at the same time obtain a better understanding of the needs they have and the challenges they face.
- Being updated on domains outside our core business, including learning about potentially interesting trends and initiatives.

SERVICES

SERVICES RESUME for SINTEF & RINVE

Project Development: Project development involves essential competences for RTOs like SINTEF. Our success and longevity rely on having the ability to create an initiative around an idea or set of goals, to create a plan for how these goals can be achieved, and to secure and coordinate the required resources. A key to successful project development is to fully understand the needs of the stakeholders and respect the rules and limitations of the ecosystem they operate in. If SINTEF is a partner in an initiative, SINTEF will typically take on the project development and project leadership roles unless there are other partners that are willing and better suited.

Community building: SINTEF leads a Norwegian national network on I&M robotics called RINVE (Robotics for Inspection and Maintenance). RINVE was initiated by SINTEF, co-funded by the Research Council of Norway, and has around 30 members. The members are mostly end users (asset owners), technology providers, service providers, research institutes and universities. The purpose of RINVE is to enable Norwegian organizations to capitalize on the opportunities within I&M robotics. A main service of RINVE is to provide innovation workshops where end users can initiate collaborations with providers and research organizations to solve key problems in their industrial domain.

Contract research: Every year SINTEF carries out thousands of research projects together with national and international clients in both the private and public sectors. The research generates direct value to the clients, but also has a societal component. To the extent possible, SINTEF strives to publish non-confidential results for the benefit of all.

Strategic RDI: SINTEF invests financial surplus into research, development, and innovation (RDI) activities that we view as strategically important. Furthermore, SINTEF participates in several virtual research centers (centers for research-based innovation), dedicated to developing the long-term competences that are needed to achieve critical innovation and value generation. The work in these centers is a collaboration between RDI-active businesses and relevant national and international research groups, with funding coming the Norwegian Research Council (a public entity which funds RDI in Norway), the industrial partners and in-kind contributions from the research groups (RTOs, universities).

Provision of technological infrastructure: SINTEF offers world-leading laboratory and test facilities in a wide range of technological fields. The advanced laboratories range from microelectronics and nanoscale technologies to the world's largest laboratory for the multiphase transport of oil and gas. Together with the Norwegian University for Science and Technology (NTNU), SINTEF is currently building an ocean technology center to make SINTEF a full-service supplier of maritime laboratory services.

TECHNICAL UNIVERSITY OF KOŠICE (TUKE)

TECHNICAL UNIVERSITY OF KOŠICE (TUKE)

Website	www.tuke.sk
Email	ladislav.vargovcik@tuke.sk lambertus.a.w.van.vonderen@tuke.sk
Phone	+421 55 602 3365
Address	TUKE, Letná 1/9 Košice-Sever, 042 00 Slovakia

ABOUT

The Technical University of Košice was founded in 1952, but its roots must be sought much deeper in the past. As early as 1657 the Universitas Cassoviensis was established in Košice. In 1762, technical education in Slovakia was elevated to a higher - education level, when the Austro-Hungarian monarch Maria Theresa established the Mining Academy in Banská Štiavnica. Besides education, this also promoted research activities in a group of scientific disciplines ranging from ore mining and processing to development and production of mining machines and technology.

Based on this tradition and knowledge in 1952 the Košice Technical College came into existence, when the Czechoslovak Government issued Directive No.30/1952 Statutes setting up three faculties, namely the Faculties of Heavy Engineering, Mining and Metallurgy. These were joined in 1969 by the Faculty of Electrical Engineering and in 1978 by the Faculty of Civil Engineering. The College was renamed into the Technical University of Košice on 13th February 1991.

Nowadays the Technical University of Košice has 9 faculties which offer a wide range of academic education and research for the East Slovak region and beyond.

The Technical University of Košice has traditionally been a centre of robotics research and innovation. Several different faculties, centres and laboratories, with various specializations related to robotics, AI/data & cybersecurity cooperate and join forces on research, development and innovation. TUKE was an academic centre of robotic research in former Czechoslovakia with the first prototype of industrial robot HYMR50 built in 1976 by the team of prof. Buda.

A typical technology centre in this framework is the Prototyping and Innovation Centre. This centre has extensive facilities for developing, manufacturing, and testing of innovations. The centre offers custom-made responses to the needs of business partners / customers.

The Prototyping and Innovation Centre also houses the Robotics Hub Kosice.

Figure 48: TUKE facilities (I)

DIH INFORMATION

- Robotics Hub Kosice serves its clients with joint research, technology transfer, development of new robots, system design/simulation, prototyping, testing under relevant conditions, and training courses and events.
- Our teams are flexibly composed, based on the needs of the project. The university and our centre offer many facilities, labs and expertise to support the work of our teams in fulfilling the project objectives.
- We are connected to several European networks of Robotic Hubs, striving to co-create new productive solutions, share our facilities and boost digitalization within our regions.
- In collaboration with industry, we have successfully completed projects in several different domains (CERN, nuclear power, automotive, aerospace, etc.).
- Industries: Manufacturing, decommissioning nuclear power plants, inspection & maintenance, agriculture.

Figure 49: TUKE facilities (II)

The RIMA network provided us with many opportunities to expand our activities and bring them to higher levels. Our contacts within the field of robotics inspection and maintenance substantially increased and/or intensified. We gained much broader and deeper knowledge about robotics applied in inspection and maintenance domains.

We learned a lot more about the RIMA partners organizations, approaches and best practices. Finally, we got new insights and experiences regarding project development within the EU funding programmes.

SERVICES AND EXPERTISE

We provide the full scale of services in terms of Technology-, Business- and Ecosystem Services like contract research, skills & training, community building etc.

The activities of our services are often related to:

- prototyping, testing and validation
- production of complex products
- design of decommissioning systems for Nuclear PP
- mentoring and advisory
- robotics
- smart manufacturing

Our specialism encompasses e.g.:

- robots for decommissioning of steam generators (Nuclear PP)
- inspection robots in radiation-contaminated areas
- zero-defect manufacturing
- intelligent robots interacting safely & helpfully with humans
- enhanced robots or automated systems with smart capabilities

Within our service range, we would like to highlight our capacity as a full-service design centre. Full-service: for customers from start-ups to large companies, from feasibility studies to systems testing & certification, from one series production to small scale volumes. Also, it is possible to pick a single service out of the service range—for example systems modelling & simulation or prototyping.

Full-service range:

- contract R&D
- feasibility studies & proof of concept
- systems modelling & simulation
- prototyping
- (pilot) system assembly & quality control
- systems testing & certification / hand-over
- service & technical support

Example: for a large company in Slovakia we performed the full-service range for a complex robotics system deployed for the decommissioning of contaminated steam generators.

I-BOTICS: TNO & UNIVERSITY OF TWENTE

TNO

Website	https://www.tno.nl
Email	tycho.brug@tno.nl
Phone	+31 88.866.0000
Address	Anna van Buerenplein 1 The Hague NL-2595DA The Netherlands

ABOUT

The Netherlands organisation for applied scientific research (TNO) is a private, fully independent, not-for-profit research & technology organization established by Dutch Law in 1932. As an organization regulated by public law, we are independent: not part of any government, university or company.

Over 3,400 TNO employees are dedicated to innovation through applied research to enhance the wellbeing of society and the competitiveness of companies. TNO has an extremely broad expertise base. TNO connects people and knowledge to create innovations that boost the sustainable competitive strength of industry and wellbeing of society. This is our mission and it is what drives us in our work every day.

In the field of robotics, TNO has extensive knowledge about telepresence, human factors, human systems integration, multisensory information presentation, brain-machine interfaces, sensor fusion, interpretation of environments, SLAM, network technologies, wearable robotics, social robotics, autonomous and semi-autonomous systems, and human-autonomy teaming.

As TNO, we can only strengthen the knowledge base for the Netherlands by working together with internationally leading knowledge partners, companies and governments. Given the globalization of research, TNO is attaining an increasingly international character.

DIH INFORMATION

TNO is in the same i-Botics DIH as UT. This DIH and the services that it offers are described below.

I-BOTICS: TNO & UNIVERSITY OF TWENTE

UNIVERSITY OF TWENTE (UT)

Website	https://www.utwente.nl
Email	i.bante@utwente.nl
Phone	+31 534 893 998
Address	Drienerlolaan 5 Enschede 7522 NB The Netherlands

ABOUT

The University of Twente is a research university with a focus on technical developments and their significance for people and society. This is reflected in our bachelor and master programmes, which are organized into five faculties:

- Engineering Technology (ET)
- Electrical Engineering, Mathematics & Computer Science (EEMCS)
- Applied Sciences (TNW)
- Behavioral, Management & Social Sciences (BMS)
- Geo-Information Sciences and Earth Observation (ITC)

The University of Twente is one of the four technical universities of the Netherlands. Originally founded in 1961 as the Technische Hogeschool Twente, it was renamed University of Twente in 1986. It offers a wide range of Bachelor and Master degrees, mainly in technical fields like physics, chemistry, mathematics, robotics, mechanical engineering, etc. From a research perspective, we believe technology and people have a leading role to play in providing solutions for complex societal challenges worldwide. Central to our vision is our High-Tech Human Touch philosophy, in which we fuse natural science, data-science, biology and engineering with social and behavioural science while maintaining a relentless focus on societal impact. Furthermore, the over 1,000 start-ups and spin-offs created by UT in its lifetime make UT the most entrepreneurial university in the Netherlands by far. Novel-T, the valorisation institute of the university, provides students and researchers with all the support and tools to bring ground-breaking research results and promising business cases closer to markets and society.

DIH INFORMATION

TNO and the University of Twente proudly initiated and jointly collaborate in innovation hub and DIH i-Botics. It develops knowledge and technology for value-adding robotic solutions. i-Botics cooperates with industrial partners in the full value chain and with government organizations. This involvement is crucial for realization of value-adding developments and successful implementation of these solutions. i-Botics is mainly active in the domains of telerobotics (haptic feedback, remote operator, remote sensing etc.), human-robot teaming and interaction, inspection & maintenance robotics, subsea robotics, and wearable robotics.

Knowledge is aggregated from various domains and researchers from our partners work together with us. It is our joint goal to develop robotic systems with enhanced situational awareness and intuitive control and thereby maximize effectiveness and efficiency.

Extensive experience in various domains and durable cooperation with partners assures successful robotic applications and creates value-adding solutions that are win-win for the stakeholders involved.

SERVICES

SERVICES RESUME for i-Botics
Business Services: Access to finance; Incubator/accelerator support; Offering housing; Project development; Skills and education. **Ecosystem Services:** Community building; Ecosystem learning; Strategy development. **Technology services**: Contract Research; Provision of technology; infrastructure; Strategic RDI; Technical support.

VTT & SMACC

VTT TECHNICAL RESEARCH CENTRE OF FINLAND

Website	https://www.vttresearch.com/ https://www.smacc.fi
Email	kirjaamo@vtt.fi juha-pekka.anttila@vtt.fi jari.erkkila@tamlink.fi markus.aho@tuni.fi kai.hamalainen@tuni.fi
Phone	+358 40 825 4229
Address	P. O. Box 1000, FI-02044 VTT Finland Kampusareena, Korkeakoulunkatu 7 1.krs, Tampere, Finland

ABOUT

VTT, Technical Research Centre of Finland, is one of Europe's leading research institutions and the major recipient of EU research funding in Finland. We are a public-sector research centre working under the Ministry of Economic Affairs and Employment. Our task is to advance the utilization and commercialization of research and technology in commerce and society.

Key figures:

- 244 total revenue, MEUR
- 2,129 employees
- 1942 year of VTT's establishment
- 440 patent families
- 27% of Finnish innovations are completely or in part results of VTT's expertise
- 45% of our turnover comes from abroad

The special duty of VTT as an independent and impartial research center is to promote the wide-ranging utilization and commercialization of research and technology in commerce and society. VTT is a non-profit company.

The VTT and SMACC (Smart machines and manufacturing competence centre) DIH gathers several legal entities: VTT, Tampere University (TUNI), Tampere University of Applied Sciences (TAMK), Tamlink ltd. Tampere University and Tampere University of Applied Sciences have just unified several research and education institutes of Tampere region. The DIH will also expand in the future, when all the stakeholders of each institute are involved to the DIH operation. Each entity has its own domain of competences and work together to stimulate innovation. All the entities are located in Tampere.

SMACC Tampere Digital Innovation Hub has laboratory facilities in many locations (and the situation is evolving):

- VTT facilities in Tampere, Espoo and Oulu.
- TUNI facilities in Tampere University, Tampere, Hervanta.
- TAMK facilities in Tampere University of Applied Sciences, Tampere, Teiskontie 33.

DIH INFORMATION

VTT has more than twenty years of expertise in robotics research. Robotics research at VTT is divided into three locations: VTT Espoo specializes in Industrial robot safety and robotics in indoor logistics and drones. VTT Tampere specializes in human-centered robotics, teleoperation, and teleoperation interfaces. VTT Oulu specializes in special robotics, such as heavy mobile robotics. Most of the personnel are experts in robotics and full-time researchers in robotics from PhD students to senior researchers and project managers. The unit also holds senior business developers who develop technical and commercial activity and bilateral links with industrial companies.

On experience, VTT has more than 80 years of experience in cutting-edge research and science-based results. Our research, development, and innovation activities are divided into three business areas: carbon-neutral solutions, sustainable products and materials, and digital technologies. At the core of our research in 2020 was, for example, building Finland's first quantum computer.

Robotic-related topics:

1. Collaborative robotics.
2. Multi purpose mobile and service robotics.
3. Cloud robotics with edge computing.
4. Industrial 3D printing.
5. Remote operation.
6. AI for mobile machinery.
7. Virtual and augmented reality.
8. Drones.
9. Agile, flexible manufacturing and production industry.
10. Safety culture assessment and development.
11. Risk and resilience management.
12. Design in research.
13. Systems engineering.
14. Smart predictive maintenance.
15. Software engineering.
16. Industrial sensing solutions.
17. Digitalisation tools.
18. AI for industry.
19. Business development from industrial and data-based services.
20. Industrial circular economy and ecodesign.
21. Innovations and business ecosystem

VTT`s responsibilities in the RIMA project included mapping the various actors in the innovation value chain across Europe; engaging and attracting start-ups, SMEs and slightly bigger companies to apply for the experiments; offering them the services provided by the RIMA network to foster the take-up of robotic solutions; and marketing and expanding the RIMA network of DIHs.

SERVICES

The services proposed by the DIH aim to create a framework facilitating the deployment of technologies, build connections and trust, and understand the needs of industrialists in order to promote economic development.

SERVICES RESUME for VTT and SMACC
VTT`s business services include: Incubator/accelerator support; Offering housing; Project development; Skills and education **VTT´s ecosystem services include:** Community building; Ecosystem learning; Strategy development **VTT´s technology services include:** Contract research; Provision of technological infrastructure Strategic RDI; Technical support; Testing and validation

SECTION C

THE IMPLEMENTATION

CHAPTER 7

SKILLS & TRAINING

TSAMPIKOS KOUNALAKIS

The RIMA consortium recognizes that robotics I&M requires a new educational basis increasing the awareness of stakeholders and end users in the relevant technologies.

RIMA includes several activities in this direction by designing and offering training services. These activities facilitate the building of trust and acceptance of robotics technologies by European SMEs and the future workforce of I&M robotics. Training services focus on creating a robotics I&M culture that will feed innovation in the targeted sectors of energy, oil & gas, water supply, transport, civil engineering and infrastructure.

7.1. RIMA training framework

The concept behind the RIMA training framework is to increase digital skills and knowledge of robotics, especially within the I&M domain, amongst a wide range of people. From students starting out on their career path, through the current workforce who need to retrain, managers who need to understand the importance and effect of digitisation, and start-up technology development companies that need concrete information and support for getting their upcoming technology to market.

To achieve these goals, RIMA designed a network training offer which is harmonized across the member DIHs, i.e., providing customized courses in local languages while covering the regional market needs, and contains four major categories based on the stakeholders that are the target participants:

End-user staff: the aim under this training category is to increase awareness of the abilities of modern robots and associated technologies amongst end users. Special emphasis is given to people that have the decision-making power to choose whether to invest in robotics or not. The focus is around showcasing available I&M robotics technologies, how to finance such technologies and hands-on introductions to inspection technologies for different applications.

Current workforce: RIMA focuses on getting the best out of whichever technologies management has chosen for existing I&M applications. This knowledge is normally passed on successfully to new staff by current operators in "on the job" training.

Technology developers: Under this category emphasis is given on how to contribute to the pool of upcoming technologies, the creation of demonstrators, the optimization of I&M robotics applications and the innovative use of standard robotics platforms.

Upcoming workforce: RIMA acknowledges that the current generation is much more digitally aware than previous generations, but this general fluency with technology needs to be sharpened into usable knowledge in I&M technologies. Thus, RIMA promotes an education paradigm equipping the upcoming workforce with the knowledge and skills required to work on robots in I&M, utilizing the state of the art in computer science, VR and simulation tools for robotics.

7.2. Description of the courses

NB. "courses" in this context includes taught courses, interactive workshops, seminars, presentations, etc.

Courses targeting the end user and senior staff:

- Introduction to inspection and maintenance robotics: This course targets anyone who wants basic insight into underlying technologies for inspection and maintenance, e.g. sensors, drones, mobile platforms and complete solutions on the market. The course will concentrate on robotic applications of I&M within e.g. energy, water, roads, railways, etc. and present cases, success stories and examples of solutions. It also teaches how we can help companies enter the market.
- Financing robotic inspection and maintenance:

This course will uncover how you can finance the development of new robot solutions. We guide you in specific financing options available Europe-wide and from financing bodies in your local region.

Courses targeting the current work force:

- Useful inspection and maintenance robotic technologies: This course targets inspectors and repairers who are experts in their field, but who would like to be introduced to how robot technology can be used in their everyday worklife. We focus on introducing robot solutions that are already on the market, the benefits of using these, and finally help in identifying potentially useful robots in your particular domain.

Courses targeting technology developers:

- Developing robot technologies for the inspection and maintenance market: This course focuses on innovation models and tools that can accelerate the development of new robotic products and solutions. The course is based on current challenges and issues within inspection and maintenance, and which innovation techniques can be useful to ensure the development of value-creating and cost-effective robot solutions for all stakeholders.
- Optimizing your inspection and maintenance technologies:

This course is aimed at technology developers in companies, including startups that develop robot solutions for inspection and maintenance. The course guides participants in how to optimize technologies from the early stages of product development, as well as sustainable methods for further development and continuous optimization of marketed solutions.

Courses targeting the upcoming workforce:

- Working with robots in inspection and maintenance jobs: The aim of this course is to introduce and inspire students, entrepreneurs and newly trained developers / engineers on how to collaborate with robots in daily work tasks. The course introduces examples of market-ready solutions, looks at the technologies behind these solutions and finally teaches you more about how to get support for your startup.
- Upcoming technologies in inspection and maintenance:

This course focuses on early technologies that have not yet been implemented in solutions on the market, (though they may be components of research solutions), but have the potential to solve specific tasks within I&M. "Technologies" include AI, collaborative robot systems, new sensor types, and mobile platforms, for example.

- Working with VR and simulation in inspection and maintenance:

This course introduces the potential of and technologies behind VR (virtual reality), MR (mixed reality) and AR (augmented reality) and sheds light on the benefits of using these technologies for inspection and maintenance.

7.3. Where can you find these courses?

RIMA DIHs are offering local courses or talks in collaboration with local clusters, training schools and universities. Courses are offered physically (where possible) and online. The newest course offerings can be found on the RIMA homepage and RIMA Network social media. If you have any questions about the courses, please contact your local DIH representative.

CHAPTER 8

CASCADE FUNDING

ANETA GAŁĄZKA

8.1. Introduction to cascade funding

Innovation in the European Union (EU) has been presented as a clear priority for the near future, especially regarding energy transition, digitalisation, Internet of Things and robotics. SMEs and start-ups with a great potential of knowledge but lack of expertise in funding opportunities are often the main recipients of cascade funding.

Cascade funding, also known as Financial Support for Third Parties (FSTP), is a European Commission mechanism for distributing public funding in order to assist beneficiaries in the uptake or development of digital innovation.

In this section, key information on cascade funding as a potential source of funding for mainly SMEs (but also large companies, university research centres, etc.) is retrieved from the experience of the RIMA project, which was funded by the EC under the H2020 programme.

The ultimate objective here is to discuss the effectiveness and advantages of cascade funding as a feasible and efficient alternative source of innovation funding for SMEs and start-ups.

8.2. Cascade funding mechanism

Cascade funding takes place in the framework of the European Union's flagship funding programmes for research and innovation Horizon 2020 and Horizon Europe (launched in 2021). Within this frame work all innovation plans are to promote and create smart, sustainable and inclusive growth and jobs in Europe.

To achieve this, the cascade funding mechanism was designed to make the process of applying for public funding from the European Commission much easier and faster. The cascade funding scheme allows European consortia to assign part of the budget to support third parties (entities that are not part of the consortium and are not considered beneficiaries by the European Commission), for the consortium to distribute through competitive open calls for proposals in the form of prizes, vouchers or grants.

The financial support is usually *equity-free*. There is no equity or share capital assigned by any party.

The process of discovering the most promising European innovators follows some main criteria established by the European Commission:

Excellence. The proposal(s) selected for funding must demonstrate high quality in the context of the topics and criteria set out in the call;

Transparency. Funding decisions must be based on clearly described rules and procedures defined and made public alongside the call, and all applicants should receive adequate feedback on the outcome of the evaluation of their proposals;

Fairness. All proposals submitted to a call are treated equally. They are impartially evaluated on their merits, irrespective of their origin or the identity of the applicants;

Confidentiality. All proposals and related data, knowledge and documents are treated in confidence;

Efficiency and speed. Evaluation of proposals and award of the financial support should be as rapid as possible, commensurate with maintaining the quality of the evaluation and respecting the legal framework.

Conflict of interest. The beneficiaries must ensure that the action is implemented impartially and objectively, especially during the evaluation and monitoring.

8.3. Who does cascade funding target?

Cascade funding supports digital innovators that can contribute to Europe's productivity, quality of life and competitiveness and also tackle societal and sustainability challenges. In practice, these innovators would generally consist of: entrepreneurs, start-ups, innovative SMEs, slightly bigger companies (mid-caps), universities, research & technology organisations. These different profiles are sometimes asked to apply together, for example to demonstrate a proof-of-concept. For instance, a university or RTO and an SME or mid-cap can work together on developing an industrial robotics solution and then apply with an end-user SME, thanks to which they will be able to validate the effectiveness of a technology in a real or realistic environment.

8.4. Cascade funding in RIMA

Three basic RIMA instruments for financially supporting the development and deployment of robots in I&M applications were chosen: two types of eligible action (Technology Transfer Experiments (TTEs) and Technology Demonstrators (TDs)) in the 1st open call and one type (Robotic Innovation Experiments (RIEs)) in the 2nd open call.

The duration of a TD was 6 months, dedicated solely to validating the prototype. TTEs and RIEs were allocated 14 months to develop a 'Minimum Viable Product' based in robot technologies—the eligible activities involved developing testing and validating the prototype.

The focus areas within Infrastructure Inspection and Maintenance (IIM) which were eligible for funding were: Energy Generation and Distribution, Water and Waste Processing, Road Infrastructure, Rail Infrastructure, Track and Trackside Equipment, Transport Hubs, Cities, and Nuclear Infrastructure.

The maximum amount of financial support granted to each micro-consortium selected in the first open call was up to € 300,000 in the case of TTEs and up to € 100,000 in the case of TDs. The maximum amount of financial support granted to each RIE selected in the second open call was up to €150,000 in the form of a lump sum payment. Payments were made by FundingBox using a result-oriented lump sum method as the default mechanism:

- The grant was paid after the completion of specific milestones. The mentors assessed progress and the project consortium took the final decision. Two main advantages stemming from this process can be highlighted: Such a procedure maximises the impact as the selected teams and organisations are constantly encouraged to implement their growth plan and achieve results in order to receive the full amount of funding they are entitled to.
- Paperwork is reduced to bare minimum. It means that the grantees are not required to present accounting documents (e.g. invoices) to prove the costs that are to be covered by the funding. They merely demonstrate the completion of the milestones set at the beginning of the programme.

Hence, the funded consortia were offered exclusive technical and business support to aid their development and speed up the industrialisation process. It is the benefits brought by this cooperation that are often more advantageous than the financial aspect of typical funding activities.

More information about funded projects can be found in the experimentation chapter.

8.5. Proactive advertising of the call

To maximise RIMA call reach, a few steps were taken by the RIMA partners:

- A communication toolkit with several materials was built (social media texts, banners, press releases, etc.) so that partners could adapt, translate and use them in their communication;
- Social media advertising: RIMA paid for promotion on social media to generate additional visibility on the favourite feeds of RIMA targets;
- Partners participated in specialised events (e.g., running a booth or making a presention);
- Organising Information Days and webinars as a great opportunity for potential applicants to ask more complex questions;
- Promoting the call in the newsletters of the partners' organisations;
- Creation of online communities and helpdesk on FundingBox spaces;
- Sending to-the-point email reminders.

8.6. Selection process

Identification of the challenges to be covered by the calls

The preparatory tasks for Open Calls are crucial for a successful selection process. It was greatly important that there was engagement from RIMA partners with technical knowledge to correctly build the documentation and enrich every step of the process. In order to specify the scope of the call for proposals across the six application domains, appropriate research and technology challenges were identified per domain. The applicants chose their challenge from a drop-down list.

Our selection process was transparent, equal and fair to all applicants, in compliance with EC standards. A description of each of the steps of our solid evaluation process id given below.

Step 1 Application through the FundingBox Platform

All candidates submitted their applications using a dedicated FundingBox page.

From the very beginning, applicants were informed about the transparency, fairness and equality of the whole application process in compliance with European Commission rules. Applicants were provided with extensive Open Call package documents: call announcement, Guide for Applicants [GfA] and annexes: informed consent form, processing of personal data, information sheet, challenges definition, RIMA equipment, facilities and competences, DIH services, eligible and ineligible costs).

Step 2 The eligibility check

All proposals received in this stage of the selection process, were checked by FundingBox Accelerator according to the following items:

- English language. English is the official language for the open calls. The proposal had to be in English in all its mandatory parts in order to be eligible.
- Submission system. All the proposals had to be submitted through the Open Call management tool [www.fundingbox.com].

- Submission before the deadline. Applications had to be submitted by the closing time and date published in the Open Call. The submission time recorded by the FundingBox platform was the official one. Late proposals were not admitted.
- Absence of conflict of interest. All cases of potential conflict of interest were assessed case by case.
- At least 2 SMEs or slightly bigger companies in the micro-consortium making the application.

Step 3 The independent evaluation

All the applications that passed the first eligibility check were assigned through the FundingBox platform to independent evaluators (proposed by RIMA partners). The latter then proceeded to make a qualitative assessment according to the three principles:

- Excellence: Ambition, innovation, soundness of the approach, cross-value-chain approach.
- Impact: Market opportunity, competition, commercial strategy and scalability, contribution to standardisation, environment and low-carbon economy contribution, social impact.
- Implementation: Legal entities, resources.

The independent evaluators were experts in the field of Inspection and Maintenance with a solid technical background as well as a good business vision. Aside from these primary criteria, RIMA partners also selected evaluators based on considerations related to gender balance, country representation and guaranteeing the absence of the conflicts of interest.

External evaluators consensus meeting

The objective of the consensus meeting was to decide the minimum score, the threshold and the list of those to be invited to the Jury Day. A thorough analysis was undertaken to identify any significant divergence among the scores of the different evaluators so they could change their score to obtain the consensus about the 'Ranked list of fundable projects', after which the best proposals were invited to the Jury Day.

Step 4 The jury day and the second consensus meeting

The applicants from the top-ranked proposals were invited to participate in a Jury Day. They were instructed to send a 10-minute pitch in advance to be displayed during the event. In addition, each consortium took part in a 10-minute Q&A session in front of a jury composed of RIMA consortium partners and external evaluators. Thus, experts could make a more informed assessment concerning the proposal, its execution and its impact. Just after the Jury Day, a selection committee gathered online for a second consensus meeting to make a final decision about which proposals to fund.

During this evaluation stage it was of utter importance that the selected proposals:

- Demonstrated that they would have socio-economic impact, reducing costs, saving time, reducing risks and increasing Europe's competitiveness
- Justified the amount of public funding they would receive through RIMA
- Favoured the development and improvement of the services offered by the corresponding DIH of the network
- Explained the benefits of funding their proposal, the impact it would have on the economy, on the life of people at work, on the competitiveness of Europe, and what perspective their application presented.
- Explained how the funding would be used and how the funding provided was used in an appropriate way.
- Explained how the project would contribute to the improvement and development of the technical and non-technical services provided by the corresponding DIH in the network.
- Agreed to contribute to promoting the results of RIMA at the end of the project.

Step 5 Sub-grant agreement signature

Following this last stage, the selected applicants were invited to provide formal documents confirming their SMEs or mid-cap status confirming their eligibility to join the support programme. Once this information was submitted to FundingBox, and it was reviewed and validated, the companies were able to sign the sub-grant agreement which was the official milestone for their commitment to the RIMA programme and the condition for the first payment. They could then start work on their experiment, using some of the services provided by DIHs and explained above.

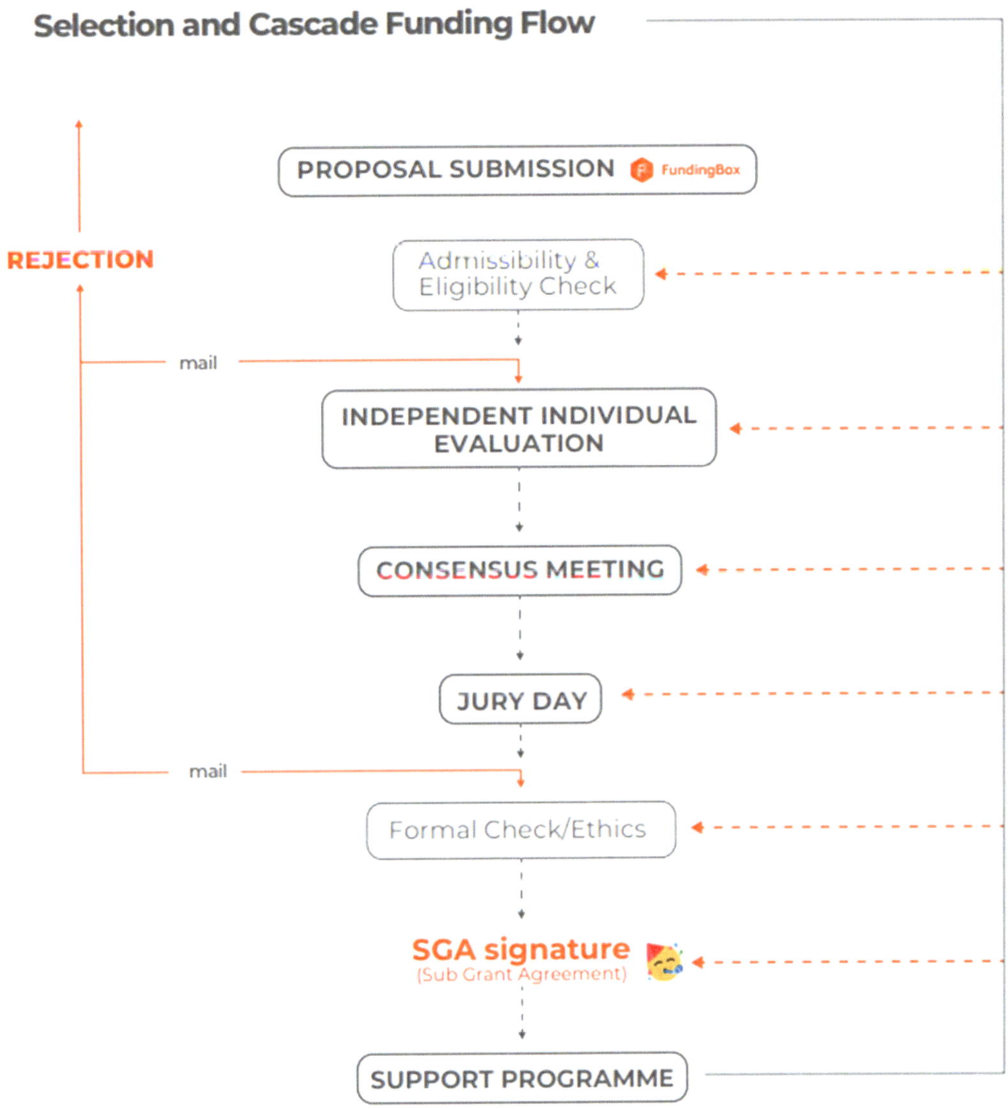

Figure 50: Selection and Cascade Funding Flow

8.7. An introduction to the support programme

Besides receiving financial support, beneficiaries were also assisted through a support programme involving technical support and business services. The aim was to ensure the sustainable success of the assets generated in the post-project phase—which ought to have been the goal for each applicant, as it contributes towards the continued existence of the company.

Together, mentors and beneficiaries determined all the aspects of the project's milestones. This includes in particular the KPIs and deliverables to take into account when assessing the team and organisations' performance.

During the course of the support programme, several interim reviews were carried out. Mentors checked the progress made by the beneficiaries and compared it to the objectives defined in the Individual Feasibility Plan. Such reviews are crucial in the framework.

8.8. RIMA Network lessons learned

APPLICATION PROCESS

- When launching the 1st Open Call, consortium partners anticipated receiving many more proposals for Technology Demonstrators (with a higher TRL level) than for Technology Transfer Experiments (lower TRL level). But in practise, the situation was reversed – partly because applicants were not thoroughly familiar with TRL descriptions. They found it difficult to place their experiment in the correct TRL category, especially given that the amount of funding for TTEs was three times more than for TDs. This caused the RIMA consortium to reformulate the RIMA instrument and the duration of the experiments. The amount of money was unified to €150,000 so as not to cause huge divergences between applications.
- Definition of beneficiary types: The 1st Open Call had a requirement for the applicants to include one technology provider and one end user as two separate companies. It turned out that there are companies that are both a technology provider and an end user at the same time. This possibility was added in the 2nd Open Call.
- During the collection of feedback, beneficiaries stated that the amount of money was not the greatest incentive, but the opportunity to take part in the programme and receive both financial and non- financial support. This was another reason to unify the amount of funding in the 2nd Open Call to €150,000 for both TTEs and TDs.

SELECTION PROCESS

- During COVID restrictions, it was proven that holding the Jury Day online had a lot of value, even if it is not an ideal solution. Online Jury Days also gave savings in travel costs, which enabled the financing of more experiments. Although there are no longer any restrictions on flying, most Jury Days were organised online for the remaining Calls.

SUPPORT PROGRAMME

- The Support Programme was extended to 14 months, which even then was not enough in some cases (mostly due to issues with obtaining the proper appliances/ equipment caused by the COVID -19 pandemic).
- There needs to be in-depth technical and administrative oversight / support available for the projects. In RIMA such support from core partners was a standard part of the cascade funding service.

GENERAL LESSONS LEARNED

- Cross-border collaboration was accomplished by the connection teams and mentors from different countries (going beyond their own comfort zone). This scheme ensured the exchange of market experience.
- The effective geographic reach (especially of countries not represented in the RIMA Consortium) of the SME interest for applying for cascade funding actions needed further investigation. In the 2nd Open Call, the satisfactory number of 6 new countries was reached.
- Cross-value-chain approach: There should be much greater emphasis on how applicants explain to what extent the proposed action will contribute to the collaboration and integration of different innovation actors, including large enterprises, SMEs, public authorities, and research organisations, across all the Infrastructure, Inspection and Maintenance sectors.

8.9. References for further information

In order to access to more information on available FSTP (Financial Support to Third Parties) calls that are open, and the history of calls launched, you can access the RIMA partner website https://fundingbox.com/moneybox2/ or the EC one here: https://webgate.ec.europa.eu/funding-tenders-opportunities/pages/viewpage.action?pageId=25559615. Moreover, you can join https://spaces.fundingbox.com/ where you can find all the information on the next calls and events.

FundingBox is one of the next-generation intermediaries allowing the money to “cascade” from European programmes to innovators (with a more than 30-year-long tradition of helping pioneer organisations obtain public funding).

It is important to know that the calls that are launched using this mechanism are very diverse in terms of verticals and technologies. Also, each of them has its own application form, eligibility criteria and financing provisions, so it’s worth analysing the ongoing funding opportunities.

To conclude, cascade funding offers numerous advantages that make it a suitable and appealing solution for new and innovative companies while in parallel enhancing collaboration and mutual learning between European stakeholders. The RIMA case study is presented as a real success story for cascade funding in an EU project framework.

CHAPTER 9

RIMA NETWORK EXPERIMENTATION RESULTS

SOTIRIS MAKRIS, GEORGE MICHALOS, PANAGIOTIS KARAGIANNIS AND SOTIRIS AIVALIOTIS

9.1. Introduction

Regular inspection and maintenance of critical infrastructure such as bridges, pipelines, and power grids are essential for ensuring the safety and reliability of these structures. Robotic technologies have emerged as a promising solution for improving the efficiency and accuracy of such operations. In this chapter, we will explore the mentoring support offered by Digital Innovation Hubs (DIHs) to small and medium-sized enterprises (SMEs) that developed 50 robotic applications funded by the 2 open calls of RIMA, focused on I&M of infrastructure.

9.2. Mentoring Activities

Robotic applications have the potential to revolutionize the I&M of different kinds of infrastructure by enabling remote and autonomous operations. These applications can perform tasks that are dangerous or difficult for humans, such as inspecting underwater pipelines or the underside of bridges or the internal walls of tunnels. Additionally, robot systems can provide large amounts of accurate and detailed data, enabling better decision-making and predictive maintenance.

DIHs play a crucial role in supporting SMEs in the development and commercialization of robotic applications. The mentoring and support services offered by DIHs include technology assessment, prototyping, testing, and access to funding opportunities. These services can help SMEs overcome technical and market-related challenges and accelerate their growth.

This chapter will provide case studies of SMEs that have benefited from DIH mentoring support in developing and commercializing their robotic technologies for infrastructure I&M in the six RIMA domains. These case studies will highlight the challenges faced by SMEs and the solutions provided by DIHs, as well as the impact of these robot technologies on the infrastructure maintenance industry. Overall, this chapter aims to showcase the potential of robotic technologies in I&M and the role of DIHs in supporting robotic development and commercialization for the benefit of society.

9.3. RIMA Network Open Call applicants

1ST OPEN CALL APPLICATIONS

The 19 funded projects of the 1st RIMA Open Call applied in two categories, namely 6 Technology Transfer Experiments (TTEs), which started from TRL 5 and aimed for TRL 7 over 14 months, and 13 Technology Demonstrators (TDs), which started from TRL 6 and aimed for TRL 7 over 6 months. The distribution of these applications was almost even among the 6 RIMA domains, namely:

- Oil & gas,
- Energy generation and distribution,
- Road, rail and infrastructure connected with cities,
- Transport hubs (ports, airports, stations etc.),
- Water supply and sanitation,

- Nuclear.

This distribution is illustrated in Figure 51. Figure 52 shows the distribution among the countries.

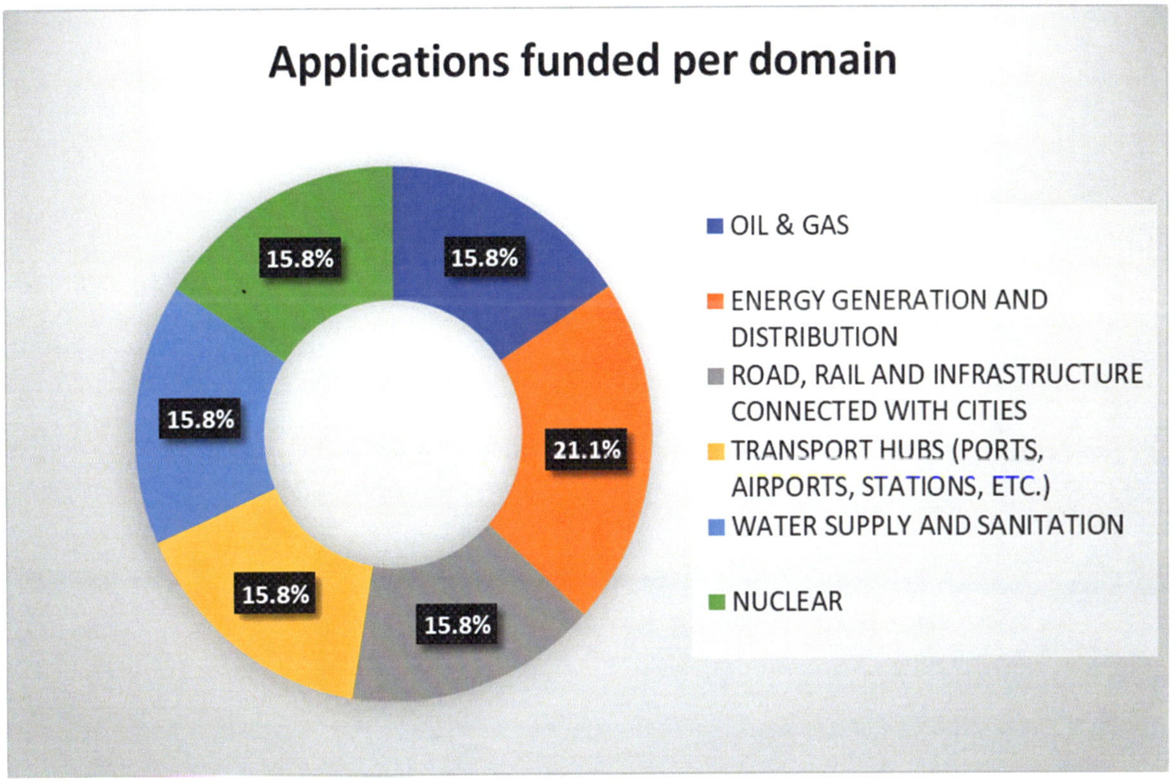

Figure 51: First call projects: distribution among the RIMA domains

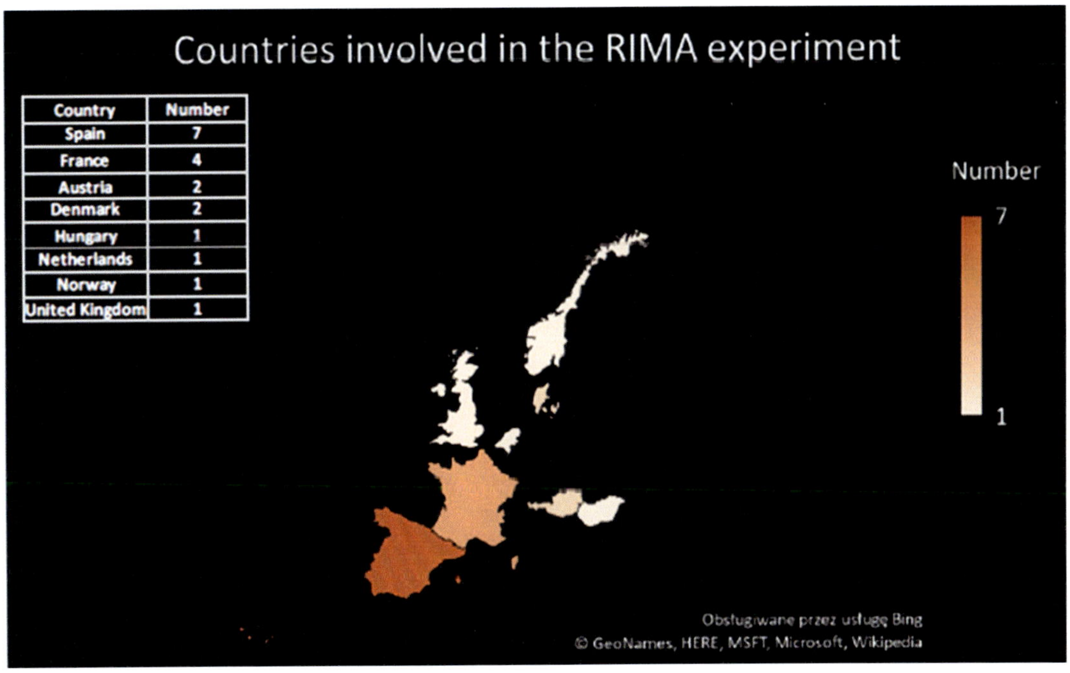

Figure 52: First call projects: distribution among the European countries

Table 4 and Table 5 below provide lists of the applications that were selected for funding under the two categories of TTE and TD respectively. The domain and a short description of each funded application is also given, along with the country where their coordinator is established.

Project Name	Domain	Tagline	Country
PRIMUS	NUCLEAR	A robotic platform enabling flexible, accurate, and cost-effective weld inspection for existing industrial infrastructure and assets.	FRANCE
WISE-INSPECT	WATER	Autonomous flying robots for I&M of underground and indoor water infrastructure (supply, storage, and sewage systems) inspection.	SPAIN
APIBOT	TRANSPORT HUBS	Airport pavement inspection based on an aerial robot for night-time data collection & AI techniques to categorize defects & calculate overall condition.	SPAIN
TANGO	ENERGY	Replacing manned helicopters by ultra-long-range drones with 90% reduced cost for energy linear infrastructure inspection (power lines).	SPAIN
ATEXDRONE	OIL & GAS	Pneumatic drone rated for explosive atmospheres for inspection of oil & gas facilities.	SPAIN
3Rc	ROAD, RAIL & CITY	Road maintenance of the future.	FRANCE

Table 4: Outline of successful TTE applications in the first open call

Project Name	Domain	Tagline	Country
UCODE	NUCLEAR	Adapting cobots to be used underwater for nuclear decommissioning operations.	UK
MIRoS		No more worries about investigation in a radiative environment, send MIRoS, a robot with complete autonomy.	FRANCE
pondsurvey.com	WATER	Autonomous industrial robot inspection and monitoring of pools.	HUNGARY
INOWATT		Safer and more efficient inspection of water treatment plants using drones.	NORWAY
RAHIP	TRANSPORT HUBS	Robotic harvesting of surplus macro algae in ports.	DENMARK
FIDR		An intelligent solution using drones and ground robots for fence inspection to optimize security and safety in all-weather conditions.	DENMARK
AeroDefectScan	ENERGY	Automated inspection system using Deep Learning (AI) to process aerial thermal images from solar plants to detect defects in solar panels.	SPAIN
DRONE4PV		Process automation for solar panel inspection with drones to decrease costs and increase the reliability of failure detection.	SPAIN
LeaksBuster		The first ROV-based solution for acoustic leak detection, which reduces inspection time by 2, increases detection rate and is easy-to-use.	FRANCE
APRORAP	OIL & GAS	Demonstration of adaptive paint stripping (blasting) robot with efficient path planning to make robotized solutions more competitive.	NETHERLANDS
RCU		Robot manipulator for inspecting complex weld geometries using phased array ultrasonics.	AUSTRIA
MAP4CAT	ROAD, RAIL & CITY	A tandem mapping system using a car plus a follow-me drone for 3D road reconstruction, with high operational efficiency, high accuracy, and low cost.	SPAIN
PreventaBot		First system to enable automated, preventative maintenance of drainage in tunnels	AUSTRIA

Table 5: Outline of successful TD applications in the first open call

More information about each project funded in the first open call follows—a short description made by RIMA personnel and the public poster made during the funded work. These details are organised alphabetically by project title.

AERODEFECTSCAN

AERODEFECTSCAN was an application submitted in the energy generation and distribution domain. Its main goal was to automate the process of inspecting solar panels using an unmanned aerial vehicle equipped with a thermal camera. This drone flies over the solar plant, capturing information and automatically processing it using AI algorithms in a ground station. The final result would be a diagnosis report with extremely valuable information for the maintenance team. The whole process after the remote-controlled flight is completely automated, so the diagnostic time and cost is low, and plant efficiency is increased. The system can be used in other sectors where thermal or non-thermal images are useful (agriculture, hydric resources, mining, building, etc.) with only with minor modifications and training of a new neural network.

Figure 53: AERODEFECTSCAN public poster

APIBOT

APIBOT, an application submitted under the transport hubs (ports, airports, stations, etc.) domain, detects defects on the airport "pavements" (runways, taxiways, apron) using camera-equipped drones. This creates a fast and non-intrusive aerial inspection solution which aims to obtain all the necessary information about defects in the airport pavement. This information was used to classify the defects visible in the images in order to calculate a "pavement condition index" (PCI). This PCI is a parameter defined by international standard ASTM D5430 and indicates the general condition of the pavement. The benefits are faster extraction of results, minimization of human presence in dangerous areas, and elimination of time conflicts with the other airport operations.

Figure 54: APIBOT public poster

APRORAP

APRORAP, an oil & gas domain application, focuses on the demonstration of a paint-stripping robot with line following and efficient path-planning capabilities. In order to reduce manual work in hazardous environments, we aim to validate a next-generation paint-stripping robot which is more efficient and effective that the mentioned manual work. This saves time and money while reducing emissions at the same time.

Figure 55: APRORAP public poster

ATEXDRONE

ATEXDRONE was an application submitted in the oil & gas domain. Its aim was to develop a pneumatic drone-based solution for inspection tasks in explosive atmosphere conditions, test it in an explosive atmosphere, obtain a standard ATEX 0/1 approval certificate, perform a first commercial reference, and start commercializing it. The benefit from using such technologies is that it becomes unnecessary for human inspectors to enter the explosive atmosphere or for the company to vent the area and this would reduce the stop time—with great benefit for the environment, worker health, and sustainability.

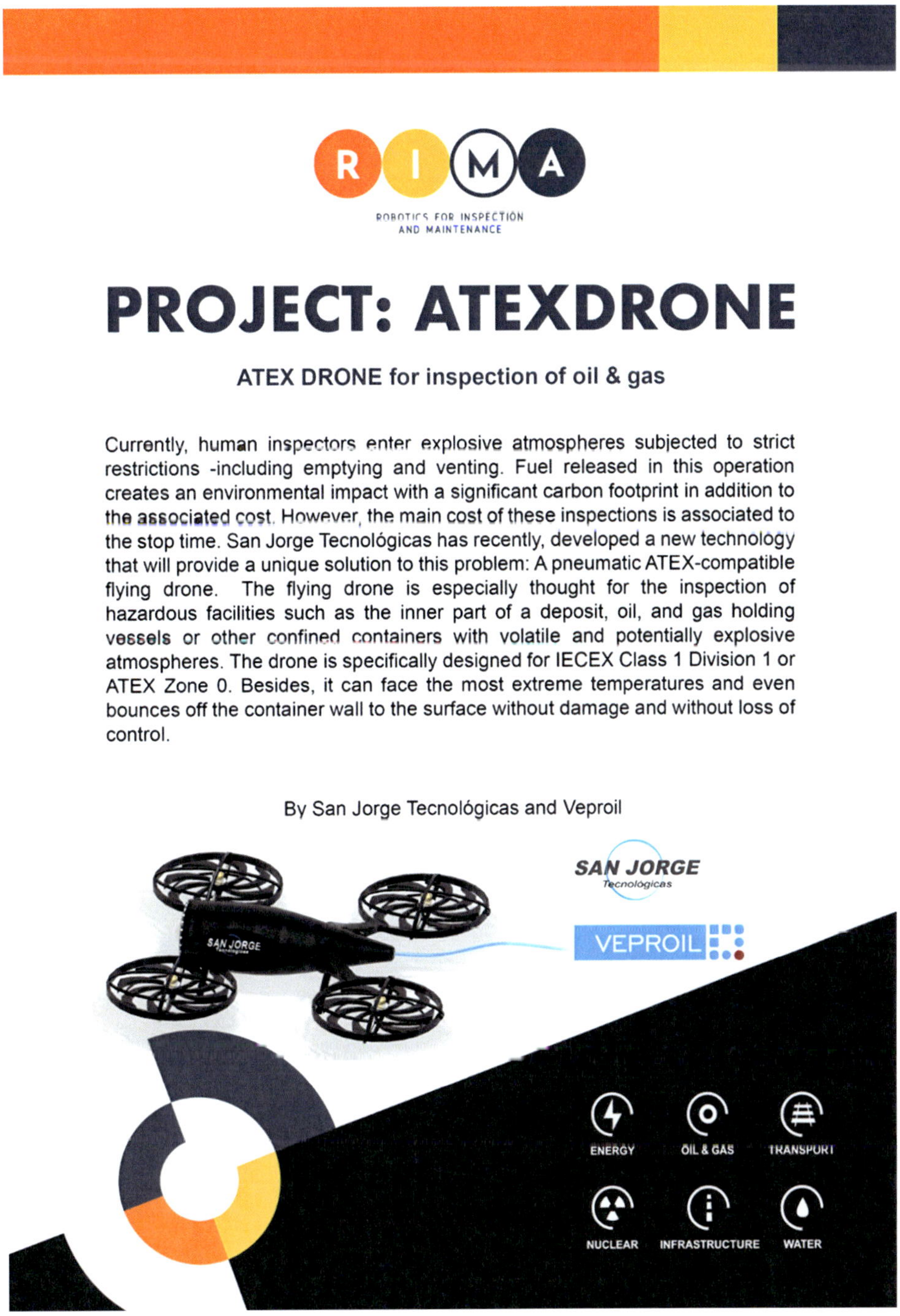

Figure 56: ATEXDRONE public poster

DRONE4PV

This energy generation and distribution application attempted to combine two state-of-the-art technologies, namely drones and AI, to develop a new product for optimizing I&M activities at utility-scale photovoltaic (PV) plants. The system combined image processing from drones with SCADA data analysis in order to detect most of the main problems at PV modules and enhance PV plant performance and reliability. Combining data analytics and flight optimization based on previous identification of potential faults and underperformance problems is a key functionality.

Figure 57: DRONE4PV public poster

FIDR

FIDR was an application for the transport hubs domain (ports, airports, stations, etc.). It focused on the development of an intelligent robot-based solution for fence inspection and automatic detection of breaches in airport and port perimeters. The custom mobile robot platform is equipped with a monochrome sensor with infrared flash while a computer vision algorithm enhanced with AI capabilities was used to find abnormalities and detect breaches. This approach enabled faster scaling of drone technology for such operations, improving their efficiency and effectiveness.

Figure 58: FIDR public poster

INOWATT

The INOWATT application was submitted in the water supply and sanitation domain. It perfomed further development of the Scout drone system, a drone able to navigate safely in GPS-denied environments due to its 3D LiDAR and state-of-the-art localization and mapping algorithms. In this way, the solution became more able to adapt to the challenges involved in the inspection of a wastewater treatment plant inside a mountain hall in Steinkjer in Central Norway. Following the RIMA project, the INOWATT team created a new spin-off company to exploit the new potential that emerged and received significant extra funding for further development.

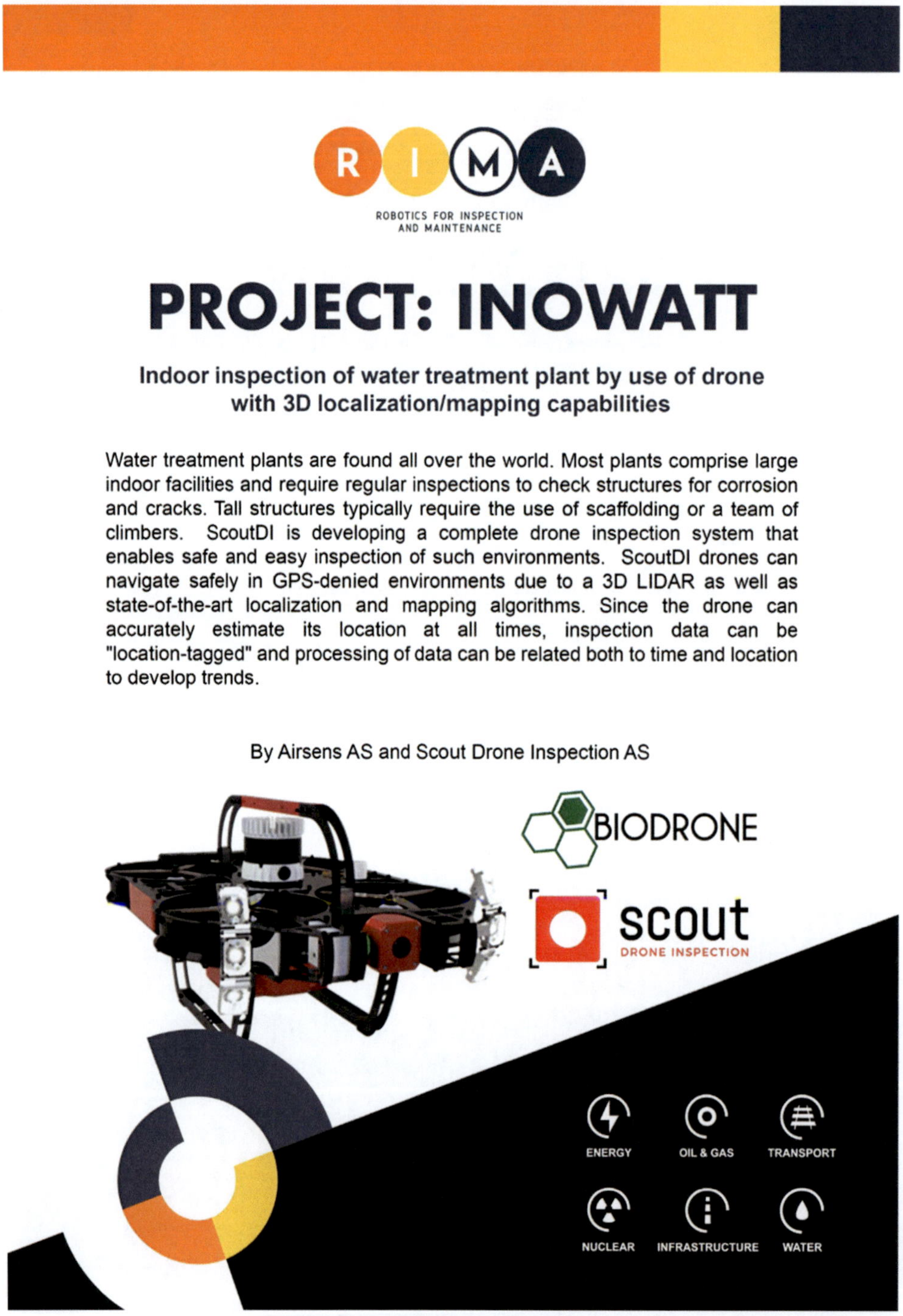

Figure 59: INOWATT public poster

LEAKS BUSTER

The project focused on the detection of leaks in canals, dams, and similar infrastructure that belongs to the Energy generation and distribution domain. The innovation in this application was that it developed the first ROV-based solution for acoustic detection of leaks. The ROV is a more reliable and affordable device that will better fit customers' needs and have a better environmental and social footprint. The maintenance of underwater infrastructure, where interventions are costly and constraining, will become easier, faster and more widespread. Jobs will be created in inspection and maintenance, the safety of dams and canals will increase, and significant water losses will be reduced.

Figure 60: LEAKS BUSTER public poster

MAP4CAT

MAP4CAT was an application from road, rail and infrastructure connected with cities domain. It used GeoNumerics' patented method for geo-data acquisition and processing to deliver an enhanced service suited for the 3D mapping of linear transport corridors (open air roads, bridges, dams) and inspection of its infrastructure (traffic lights, vertical signalling, lane separation structures, etc.). MAP4CAT brought together the two latest mapping revolutions (aerial robots and mobile mapping systems) in an innovative operational scheme: a mapping drone autonomously following a mapping car to deliver simultaneous air/ground 3D geo-data from corridors. In this way, MAP4CAT mitigates the current bottleneck connecting high-potential technology to the 3D mapping market.

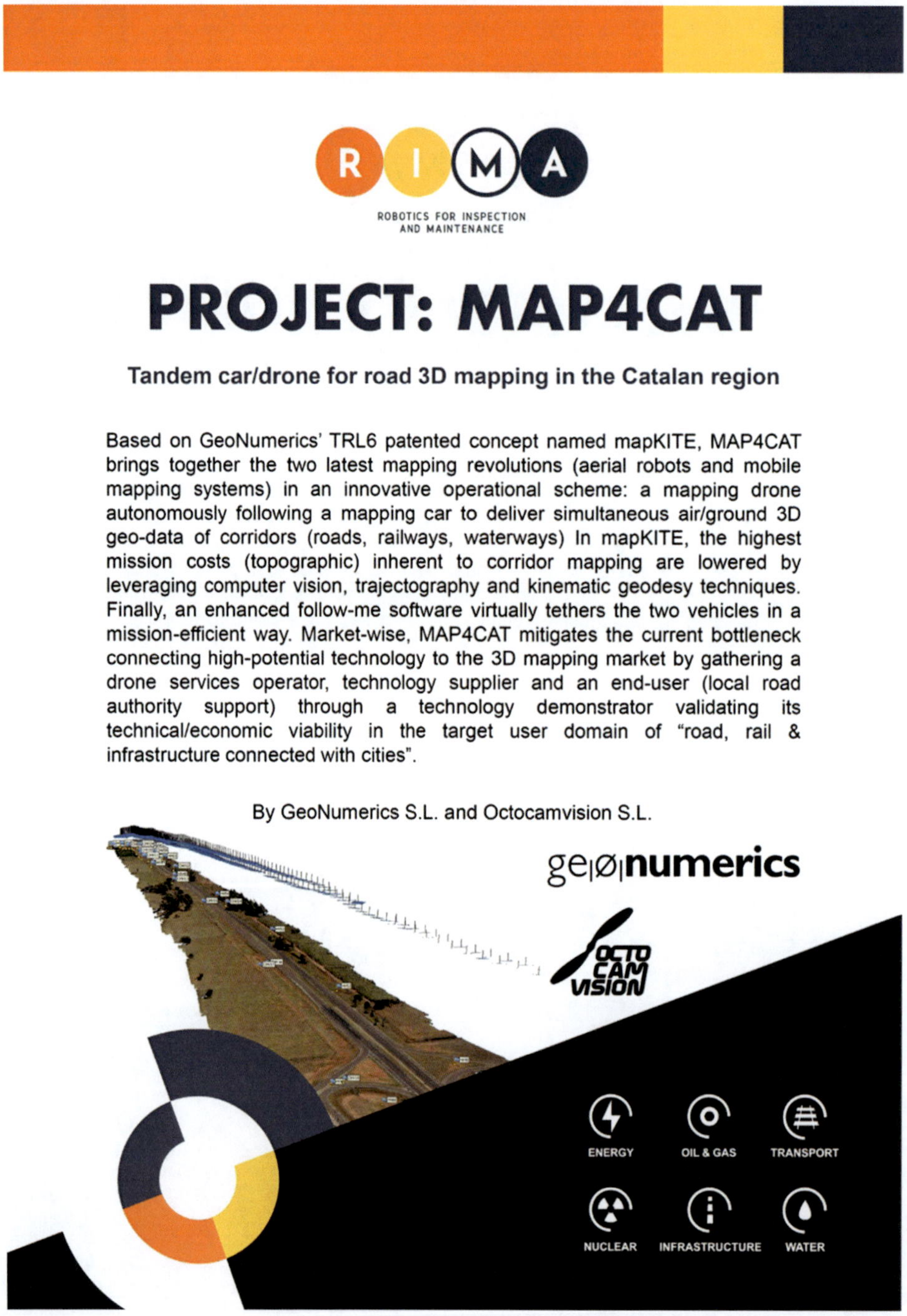

Figure 61: MAP4CAT public poster

MIROS

MIROS (Mobile Innovative Robot with Sensor) is a sensor-robot solution for nuclear facility inspection, designed and created to realize video inspection and measurement campaigns in a radiative environment. The embedded sensors are dedicated to specific investigation missions such as: visual inspection, localization & mapping, or dose-rate measurements. The integrated sensors are carefully developed/chosen based on low consumption. Wireless-operated, the embedded technologies in MIRoS enable simple, safe, and reliable investigative missions.

Figure 62: MIROS public poster

PONDSURVEY.COM

PONDSURVEY.COM was an application in the water supply and sanitation domain that focused on the development of an autonomous robot for monitoring pools. It consists of a floating platform equipped with a multitude of tools that measure different physical, chemical, biological, and other properties of the pool and its contents. Using Unmanned Surface Vehicle with mobile sensors allows facility owners to mitigate risks, comply with regulatory requirements and gather high resolution multimodal localized data that supports decision making and predictive maintenance.

Figure 63: PONDSURVEY.COM public poster

PREVENTABOT

This application targeted the road, rail and infrastructure connected with cities domain. It used a combination of robotics, electric mobility, advanced data analytics and automation to create a crawling robot equipped with cameras and brushes to clear drainage pipes. The PREVENTABOT solution focusses on the more efficient, safer and more environment-friendly maintenance of tunnel drainage systems through the combination of the above novel technologies. In this way, this solution reduces the hazard to the employees while minimising resource consumption and tunnel-locking times, minimising disruption to passengers and transported goods.

Figure 64: PREVENTABOT public poster

PRIMUS

The PRIMUS application submitted in the nuclear domain aimed to improve non-destructive testing (NDT) operations in complex and hazardous environments. The PRIMUS solution included a software and hardware package of sensors that can easily retrofit any off-the-shelf industrial robotic manipulator into a flexible, accurate, and cost-effective platform for NDT activities using advanced sensing and control technology. This solution will reduce operator exposure and increase operation efficiency, allowing NDT service providers to respond effectively to the increasing demand for industrial infrastructure and asset inspections. After RIMA, PRIMUS launched a new product and a new service in the market, as well as acquiring significant extra funding.

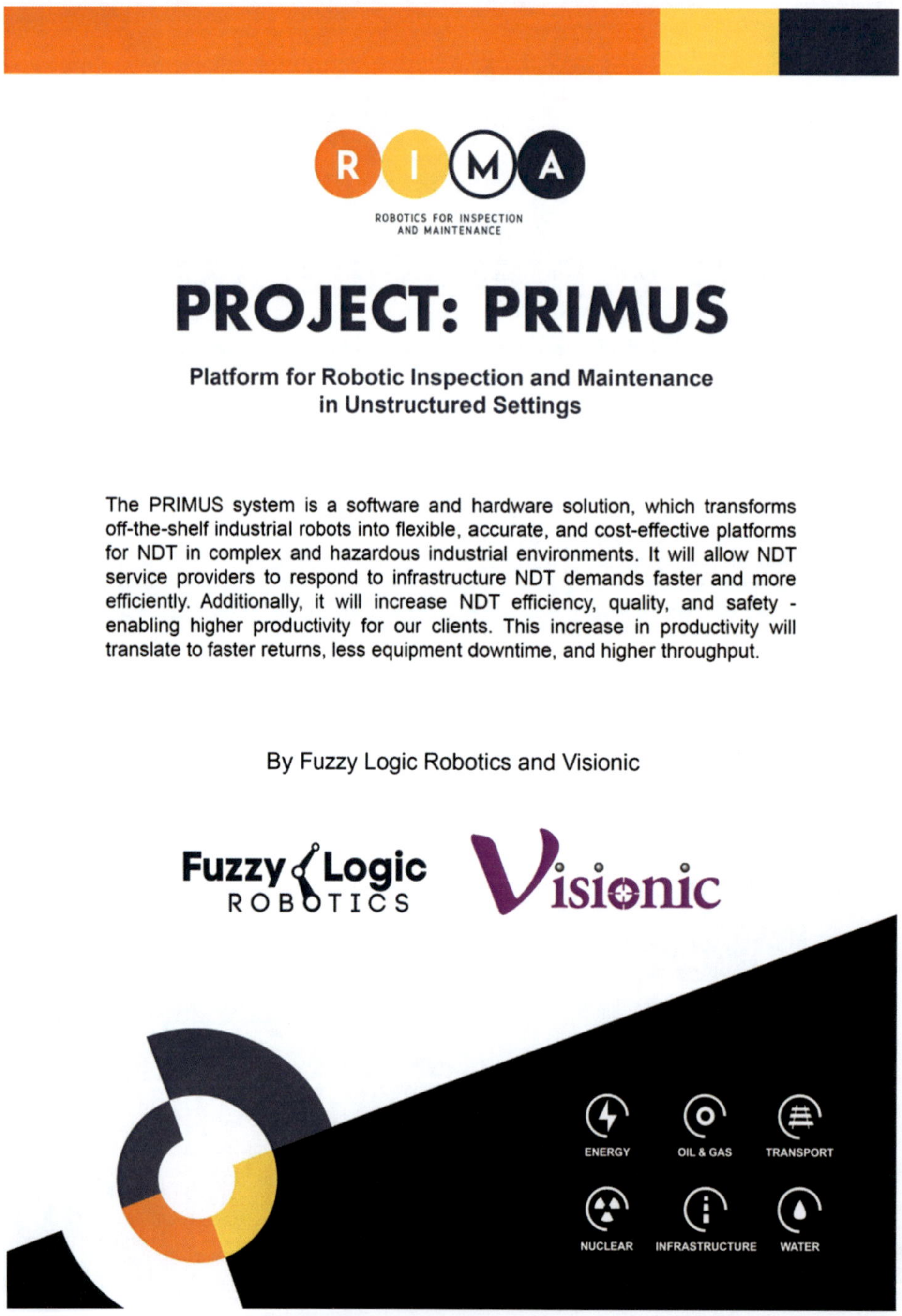

Figure 65: PRIMUS public poster

RAHIP

The RAHIP application in the transport hubs domain (ports, airports, stations, etc.) focused on the development of a robotic solution for harvesting excess algae so that they can be utilized for energy production. Algae has a high growth rate - up to 35% per year - and generally has a negative impact on the environment. A lot of money is spent on collecting, handling, and destroying algae in both the public and private sectors. The primary reason for the high cost is that the harvesting process is time-consuming, monotonous, and carries a safety risk if this process is performed manually. The benefit of the RAHIP solution is that it considerably facilitates harvesting, which increases the likelihood of establishing a production that can compete with other forms of biomass. During RIMA support, the RAHIP team managed to double their employees, while after the end of RIMA activities, the team launched 1 new product and 1 new service.

Figure 66: RAHIP public poster

RCU

RCU was a very special application for the oil & gas domain which aimed to develop a unique, custom robotic manipulator able to crawl in the pipes or any of the other complex weld geometries that are dominant in this domain. In this way, the custom robot will be able to inspect them using an Ultrasonics Phased Array sensor and integrated software for control and data display. This system will have a direct impact on the wellbeing of the workers, neighbours, and society in general as it provides tools to increase the safe operation of industry and safe use of infrastructure in multiple areas such as process manufacturing, power, bridges infrastructure and all sectors where welding is used for construction.

Figure 67: RCU public poster

RRRc

This application, submitted in the road, rail and infrastructure connected with cities domain, aimed to detect and repair cracks in the roads. RRR, also known as 3Rc, demonstrated a holistic solution that contained an industrial robot on a mobile base. The system was equipped with the necessary sensors and tools to detect the crack, move up to it, prepare the crack for repair and fill it with the pre-defined repair materials. With this solution, the team will be able to reduce the hard working conditions of the workforce through automation of some tasks. This will limit human intervention to the right level, secure the construction site, reduce repair costs and increase road lifetimes.

Figure 68: 3Rc public poster

TANGO

TANGO was one of the most innovative applications in the energy generation and distribution domain. It aimed to develop a solution capable of performing the first Beyond Radio Line Of Sight (BRLOS) drone-based power line inspection. The technology developed is a hybrid plane/multirotor aircraft with an efficient plane mode with 5 times more range than conventional drones. Additionally, this hybrid aircraft is equipped with sensors to detect defects on the power lines and avoid obstacles for a safe flight. The expected cost reduction is more than 90% compared with current solutions. During the RIMA project, the team managed to increase their employees and received significant new funding for further developments.

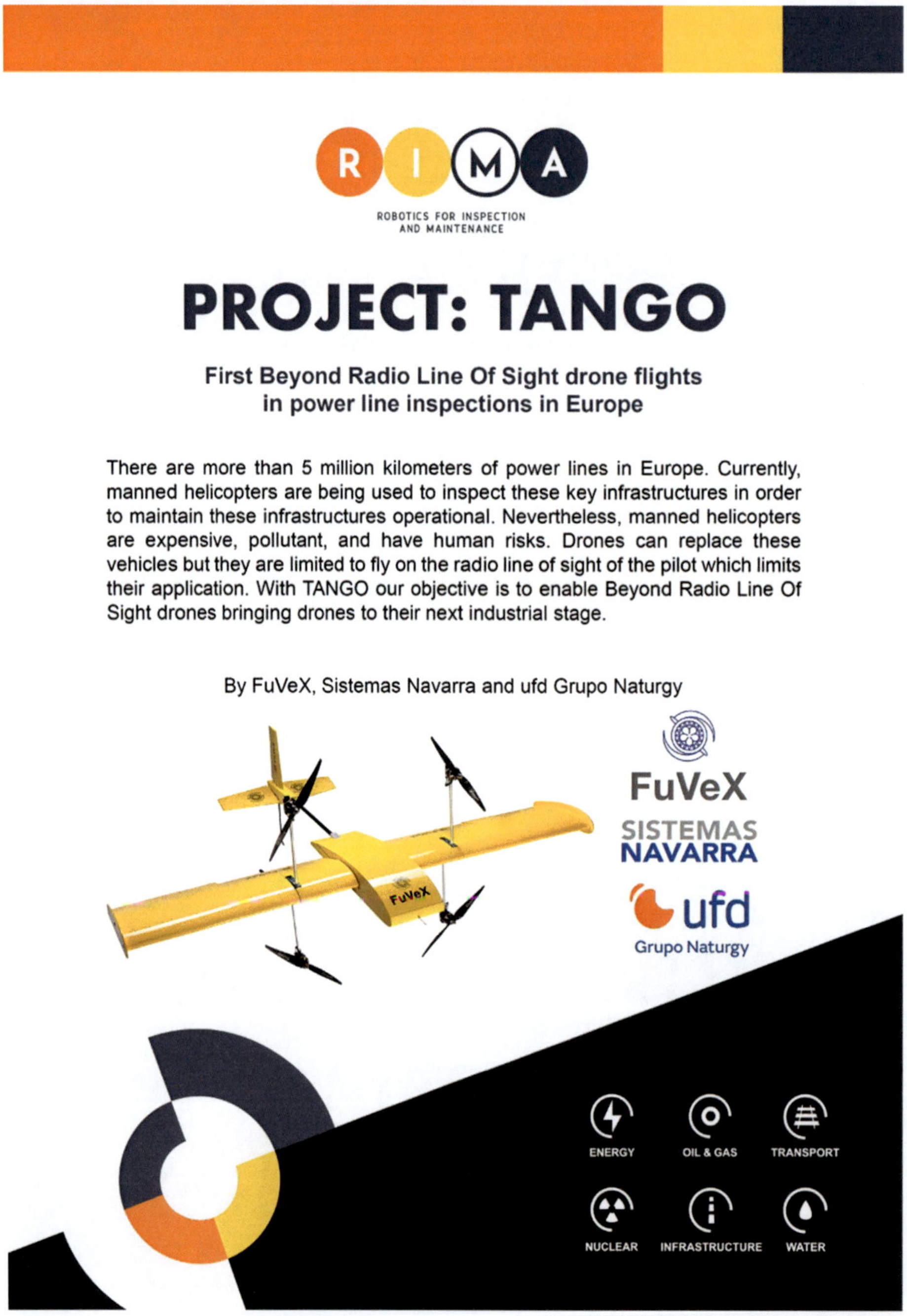

Figure 69: TANGO public poster

UCODE

This project in the nuclear domain demonstrated off-the-shelf collaborative robots for shallow water operations. Cobots are made for dexterity and operating safely in complex environments, and these features are also needed for hazardous remote operations. The project demonstrated that a cobot protected for shallow water operation is a low-cost, general-purpose solution to typical underwater operations that might be encountered while decommissioning a nuclear reactor.

Figure 70: UCODE public poster

WISE-INSPECT

WISE-INSPECT was an application in the water supply and sanitation domain of RIMA. The target was the development of a drone-based solution that could fly autonomously using ad-hoc on-board sensors that provide reliable and "zero latency" information to the drone processing core. Advanced SLAM algorithms and machine learning algorithms for self-navigation ensured the autonomous navigation of the system, while pattern matching and image analysis algorithms enabled the system to assess the condition of specific infrastructure areas. The benefit is the reduction of human intervention in the inspection of the water supply and storage systems, which is characterized by long and narrow distribution channels and storage tanks made of reinforced concrete.

Figure 71: WISE-INSPECT public poster

2ND OPEN CALL APPLICATIONS

Following the feedback received from the 1st RIMA Open Call, the application structure was slightly different in the 2nd RIMA Open Call. In this call we had one category of applications, namely Robotic Innovation Experiments (RIEs), which started at TRL 5 and aimed to reach TRL 7 during the 14-month period. The domain names and descriptions were also slightly fine-tuned for clarity, thus the domains for the second call were as follows:

- Urban and suburban transport routes and buildings
- Energy generation and distribution
- Transport, cargo and mobility
- Water supply and sanitation
- Nuclear
- Oil & gas and chemical

The outcome of the 2nd Open Call and its evaluation process was the selection of 31 RIEs. Table 6 below provides more details of the applications selected, including the domain, a short description, and the country where their coordinator is established.

Project Name	Domain	Tagline	Country
A3D	Transport, cargo and mobility	Development and field testing of an automated drone for dent detection and 3D measurement.	France
ACIDRONES	Transport, cargo and mobility	Fast and digital ship inspections using contact inspection drones.	Spain
AUPIR	Transport, cargo and mobility	Autonomous underwater quay wall visual inspection.	Denmark
NYMO	Transport, cargo and mobility	Never send a man to do a machine's work.	Estonia
RoboHop	Transport, cargo and mobility	Agile low-cost robot for dry inspection.	Greece
AutoDrone	Energy generation and distribution	Lightweight localization and mapping for autonomous navigation of an indoor inspection aerial drone.	Switzerland
BladeBUG	Energy generation and distribution	Reduce the complexity and risk of offshore wind turbine maintenance through the use of an adaptable and universal robotic system.	United Kingdom
SBID	Energy generation and distribution	Enabling fast, on-demand and safe offshore windmill inspections at reduced cost and carbon emissions via shore-based hydrogen drones.	Netherlands
SPARC I&M	Energy generation and distribution	SPARC I&M solution paves the way towards a new holistic approach for the autonomous cleaning and inspection of solar panels.	Luxembourg
AUTOBOLT	Energy generation and distribution	Development of a bolt monitoring, inspection & maintenance robot for wind turbines.	Netherlands
T-IRIS	Energy generation and distribution	We integrate inspection robots and a digital twin maintenance platform as an end-to-end solution for operators of industrial plants.	Switzerland
ROBULCLEAN	Energy generation and distribution	Robotized cleaning system through ultrasonic waves at concentrated solar power (CSP) and photovoltaic plants.	Spain
GUMBOT	Urban and suburban transport routes and buildings	An intelligent and eco-friendly solution for effective removal of chewing gum contamination from sidewalks using urban mobile robots.	Denmark
InspectRAIL	Urban and suburban transport routes and buildings	The new concept of inspecting dangerous places on railways with autonomous mobile robots will increase railway safety.	Slovenia

GreenBee	Urban and suburban transport routes and buildings	Robotic system for road infrastructure inspections.	Lithuania
AID4asBuild	Urban and suburban transport routes and buildings	Autonomous UAV platform for precise inspection of indoor building assets.	Greece
CAARMS	Urban and suburban transport routes and buildings	Connected AI-driven autonomous robotic maintenance system for road defects of civil infrastructure.	United Kingdom
CQMR	Urban and suburban transport routes and buildings	Development and demonstration of a coating quality measurement robot for preventive maintenance.	Netherlands
TunnelEye	Urban and suburban transport routes and buildings	TunnelEye enables fast, cheap and safe road tunnel inspection through a multi-sensor system with AI-based solutions for reporting defects.	Italy
RAID	Oil & gas and chemical	Real-time robotic AI defect detection & dimensioning.	Ireland
F3DI	Oil & gas and chemical	Fast-3D inspection - a marine digitalization service.	Germany
MIMIC	Oil & gas and chemical	Inspecting and maintaining industrial infrastructures remotely using the MAGOS platform, fast, with safety and accuracy.	Greece
REALIGNED	Nuclear	Robot-enhanced automated landmark inspection & geolocating for nuclear environment dismantling.	France
RoMy	Nuclear	Explore beyond limits.	France
DEMZONE-2	Nuclear	3D radiological mapping of a nuclear facility in decommissioning.	France
Magic Lancer II	Nuclear	Rad-hard electronics for use on robots.	Croatia
SPIIDER	Nuclear	The SPIIDER project will demonstrate ORANEF, a small-size and retractable indoor inspection drone for high radiation zones in nuclear plants.	France
TERRAPIN	Water supply and sanitation	Robotic underground water pipe inspector.	Greece
MOVEX	Water supply and sanitation	Remotely operated amphibious vehicle for carrying out non-destructive measurements and inspection of uncleaned sewage systems.	Belgium
PIPEYE RIE	Water supply and sanitation	Pipe inspection ROVs with light-field imaging for 3D reconstruction to cost-effectively and safely monitor submersed critical assets.	Spain
AI Water	Water supply and sanitation	Automatic leakage inspection for overhead water transportation infrastructure using aerial robot and AI computer vision.	Spain

Table 6: Outline of successful RIE applications in the second Open Call

An overview of the distribution of the different RIEs among the 6 RIMA domains is given in the graphic below (Figure 72).

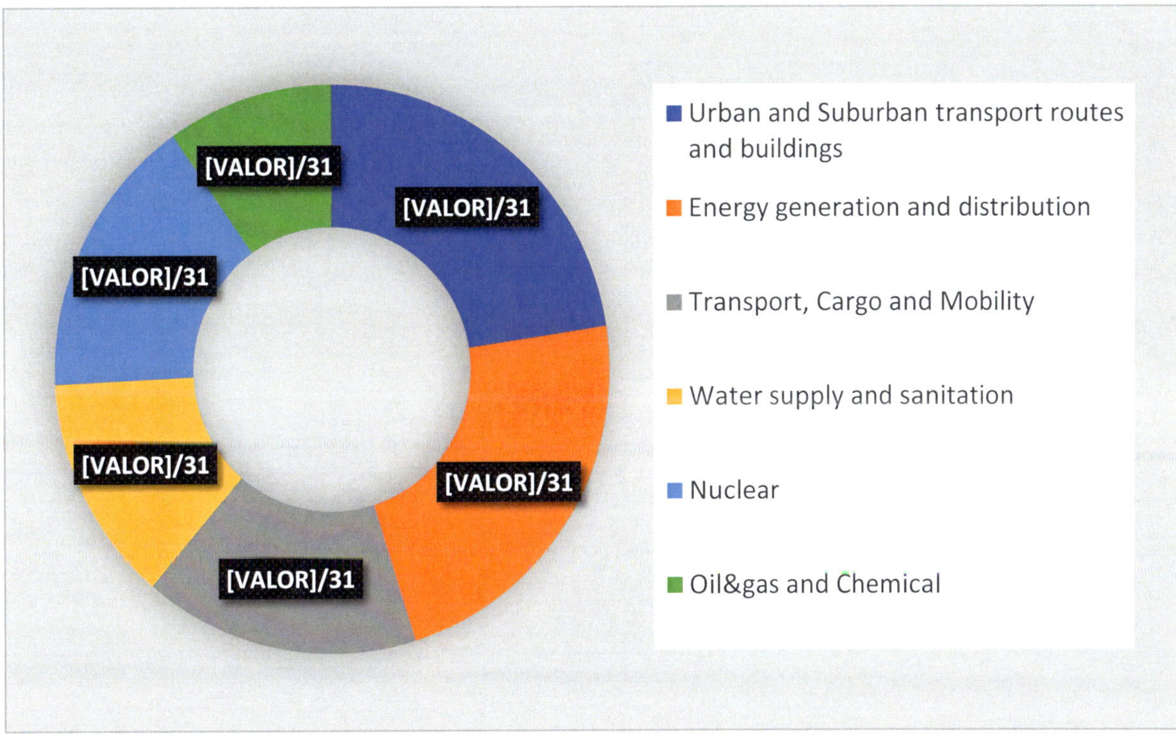

Figure 72: Second call funded projects: distribution among the RIMA domains

An overview of the countries involved in the RIEs selected in the 2nd Open Call is shown in the following figure. It shows a heatmap of Europe, highlighting the 17 countries where the 31 coordinators are located (Figure 73).

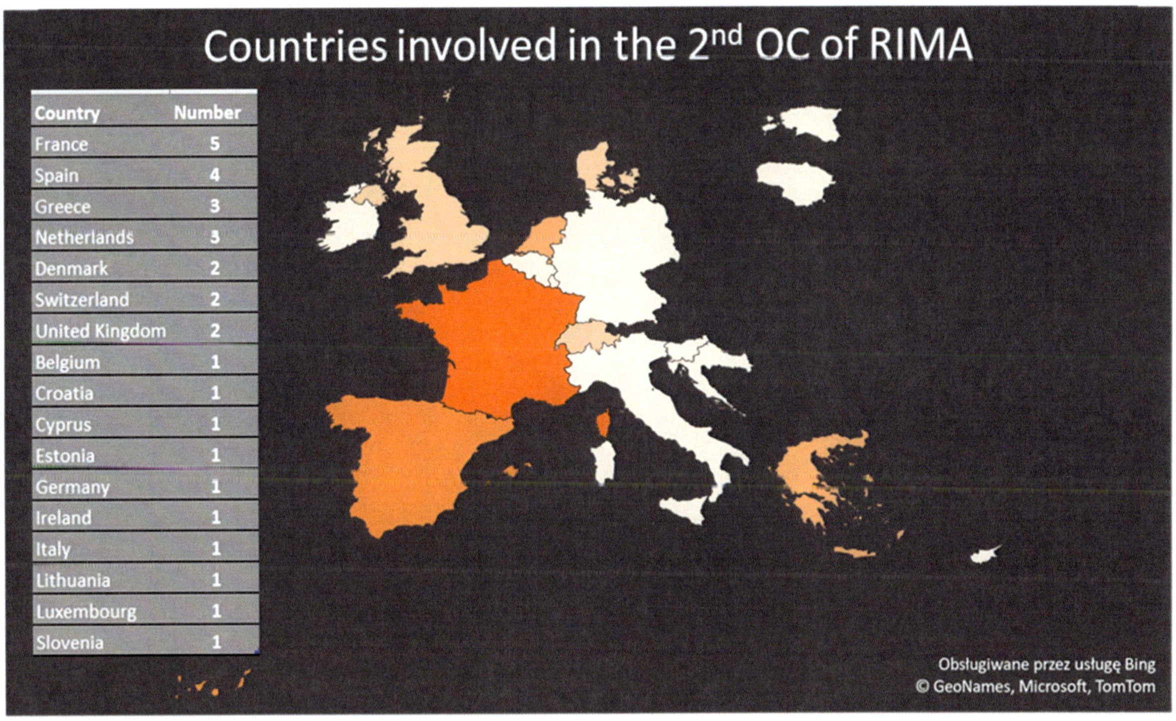

Figure 73: Second call funded projects: distribution among European countries

More information about each project funded in the first open call follows—a short description made by RIMA personnel and the public poster made during the funded work. These details are organised alphabetically by project title.

A3D

The A3D application was submitted in the transport, cargo and mobility domain. It aimed to solve the dent damage that occurs on aircraft due to loading operations, hailstorms, bird strikes, etc. The system comprises a Donecle drone equipped with a 3D scanner and is able to perform much faster inspection, which can precisely locate the dents relative to the aircraft structure. Worker safety is improved (no need to climb on upper surfaces and dangerous areas) and the process is largely paperless.

Figure 74: A3D public poster

ACIDRONES

ACIDRONES was an application submitted in the transport, cargo and mobility domain. It focused on inspection activities in the tanks of large merchant ships, such as general visual inspection, wall thickness inspection and paint coating inspection. This solution uses VES contact drones which provide data to AI-based algorithms to detect the defects. The main benefits of this solution include reduced time and cost for these operations, reduced risk for the human operators, and improved inspection quality including traceable messages.

Figure 75: ACIDRONES public poster

AI WATER

This application, as can be inferred from the name, belongs to the water supply and sanitation domain. It used an aerial robot to detect leakages in aerial water transportation infrastructure using thermal imaging and deep learning. The technologies combined are a drone equipped with a thermal camera and a GPU to process the signals in real time based on AI image processing algorithms, plus a GPS system to provide the current location. The benefit of this solution is automation of the inspection of these large facilities. Inspection is costly and hard to perform efficiently by humans alone. Lastly, the team foresees that such a solution can be easily adapted for other sectors such as agriculture.

Figure 76: AI WATER public poster

AID4asBuild

This application in the urban and suburban transport routes and buildings domain aimed to automate the inspection of building assets. The main technology used is an autonomous UAV equipped with sensors to navigate and map confined indoor spaces, collecting data that are analysed and used to create a digital twin of the building asset. One of the benefits of this solution is reduced human exposure to dangerous and confined areas, needed for surveying or inspection activities. Another benefit is the provision of precise, safe, and timely measurements through the automation processes that are stored in the cloud, accessible from anywhere and at any time.

Figure 77: AID4asBuild public poster

AUPIR

AUPIR developed an autonomous underwater inspection robot for the transport, cargo and mobility domain. It was equipped with sensors and actuators to navigate efficiently underwater while performing inspection activities on large structures. The robot is also equipped with 3D sonar and other sensors to produce a gap-less 3D model of the entire scanned area for quick maintenance. This solution will be used for underwater port structures, including quay walls, corners, round structures, and locks. The current approach requires experienced divers to collect video and experts to review hours of video to get an overview of conditions or find faults. Our system provides the possibility of automatically inspecting large areas in a stable, consistent, and highly repeatable pattern.

Figure 78: AUPIR public poster

AUTOBOLT

AUTOBOLT targeted I&M activities in the energy generation and distribution domain. It provides an innovative solution for enhanced bolt monitoring, inspection and maintenance during a wind turbine's lifetime. The custom manipulator is equipped with an ultrasonic sensor which is able to measure the pre-tension in each bolt and report this to operators, allowing them to avoid checking all the bolts using a heavy tool which measures and adjusts the torque in each nut. The benefits are that the safety and reliability of the wind turbines is increased, bolt failure is reduced, and heavy tool handling is prevented.

Figure 79: AUTOBOLT public poster

BladeBUG

BladeBUG was an application from the energy generation and distribution domain. It developed a custom, lightweight, 6-legged mobile platform, able to adjust its legs and walking gait to move around complex and curved geometries such as wind turbine blades. This robot is equipped with sensors and other equipment that enable the technicians to perform inspection, maintenance, and repair tasks on the blades. These tasks are currently performed by experts that climb and access the blades using ropes. The key benefits are the improvement in health and safety of the workers, minimizing inherent risks to personal safety, as well as minimizing the high costs of these operations. The robot has been tested in different weather conditions and shows good resilience, increasing the weather window in which inspection and repair of turbines can take place.

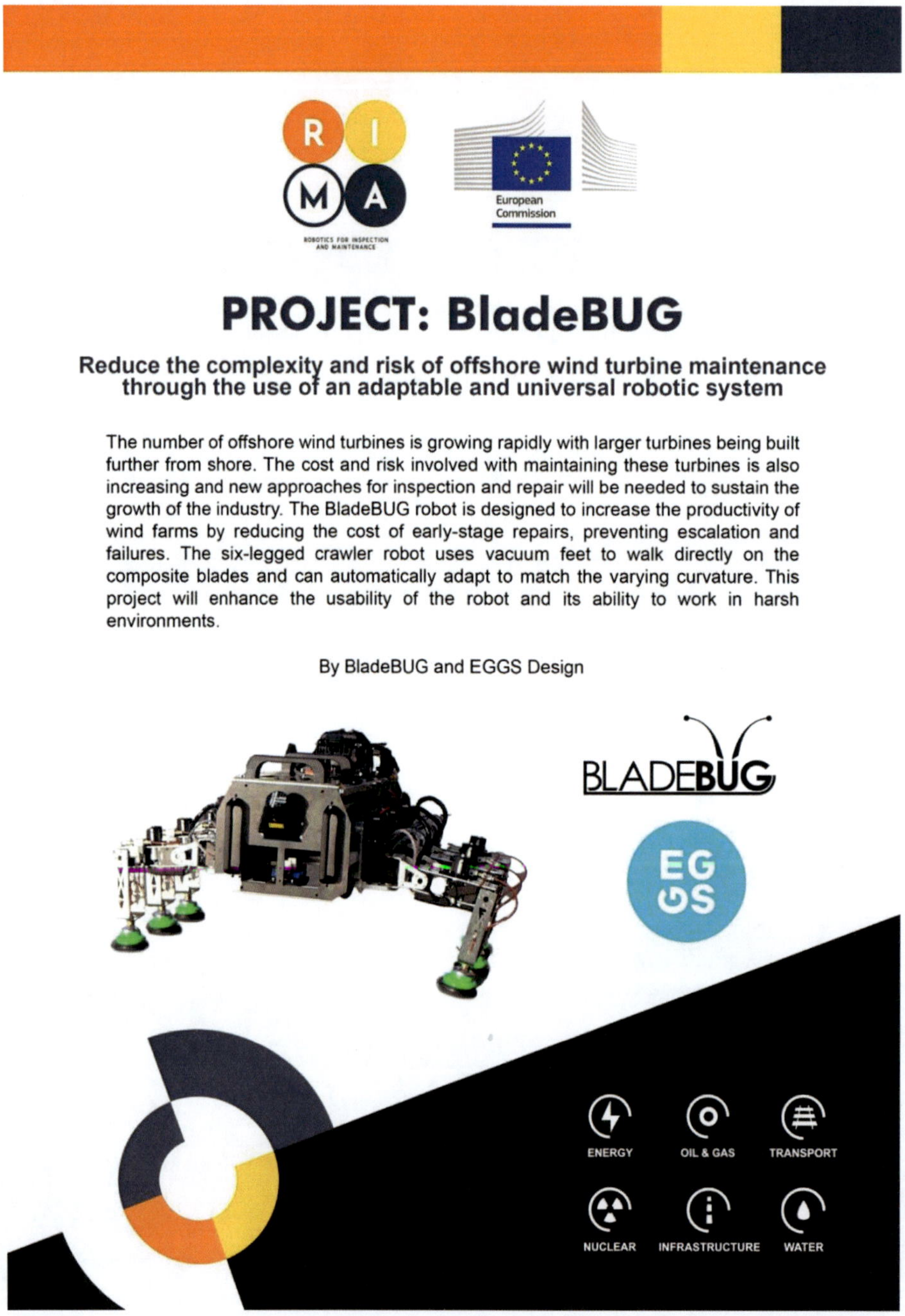

Figure 80: BladeBUG public poster

CAARMS

CAARMS was an urban and suburban transport routes and buildings application. Their system claims to be the first of its kind to detect, localize, and characterize road surface defects with volumetric reconstruction, and to predict the evolution of cracks and potholes by fusing weather and traffic data facilitated by purpose-built AI algorithms. The solution consists of a custom sensor able to detect cracks that can be attached to any mobile platform and filling system to automatically repair them. The benefit is a cheaper, faster, and safer application which is an industrial game-changer due to its autonomous, data-driven, and low-emission approach.

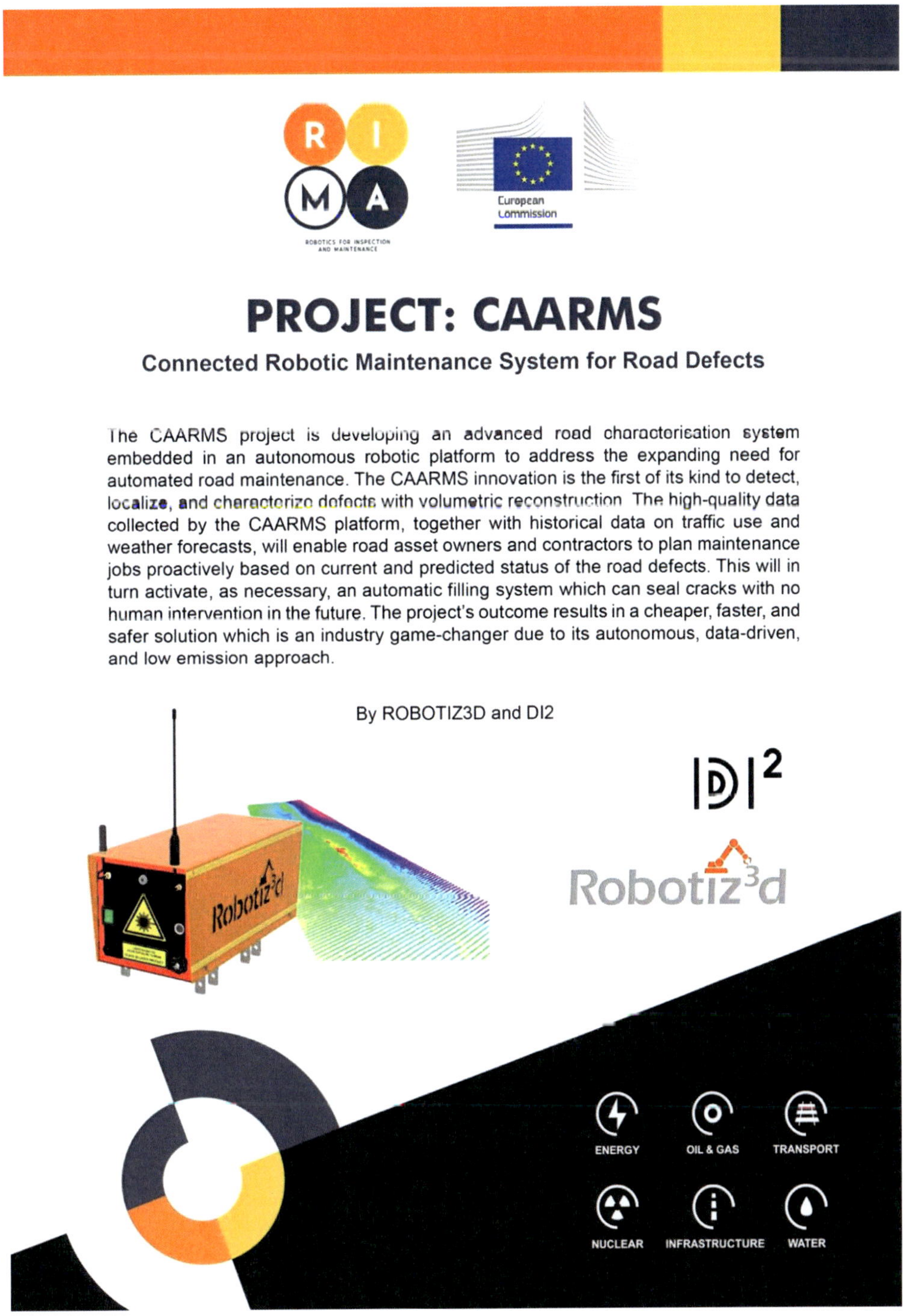

Figure 81: CAARMS public poster

CQM-R

Another application in the urban and suburban transport routes and buildings domain is that of Coating Quality Measurement Robot, namely CQM-R. This custom crawling robot is able to navigate on steel surfaces, even if they are vertical, and check for early signs of corrosion. To achieve this, the robot places magnetic pads on the surface which are removed after one to two weeks once all the necessary data has been collected. The team has developed a custom method to analyse this data and is capable of detecting the barrier qualities of the coating layer as well as its thickness. The benefit of this approach is improved prevention of corrosion using a cheaper and more thorough method than the current state of the art.

Figure 82: CQM-R public poster

DEMZONE2

DEMZONE2 was an application for the nuclear domain. It is an automated sampling device which measures the penetration thickness of radionuclides in order to obtain 3D radiological maps. The device is attached to a mobile platform so is able to perform an innovative site mapping (indoors or outdoors). This makes it possible to prioritize decontamination projects, which will have beneficial effects both on the environment and public protection, as well as enabling earlier release of potential building land.

Figure 83: DEMZONE2 public poster

GREENBEE

GREENBEE was an application from the urban and suburban transport routes and buildings. It equipped an autonomous drone with sensors able to collect data from road inspection and analyse it using AI and ML-based algorithms. The current approach is manual inspection performed periodically by specialists; thus, the benefit of the GREENBEE solution is to increase the efficiency of these inspections, minimizing the cost and intensity of human labour which could lead to human error.

Figure 84: GREENBEE public poster

GUMBOT

GUMBOT was an urban and suburban transport routes and buildings application targeting clearing one of the worst litters on the roads: chewing gum. The solution comprises a custom mobile robot that is equipped with sensors that take advantage of AI-vision based algorithms to detect the gum on the pavements plus an eco-friendly cleaning module that can guarantee a removal time of 5 seconds per gum. The benefit of this application is the low cost, cleaner streets, minimization of labour intensity, and increased task efficiency.

Figure 85: GUMBOT public poster

InspectRAIL

InspectRAIL focused on I&M activities on railway infrastructure, thus it belongs to the urban and suburban transport routes and buildings domain. The proposed technology consisted of a custom mobile robot hanging from two steel ropes outside the train profile (see picture on the poster), which would navigate the line, using LiDAR and other visual sensors to check tracks for fallen rocks, floods, and other hazards. This solution was designed to work 24/7 either on a scheduled basis or on demand, would be able to operate both day and night, and to withstand rough weather conditions. The benefits of this solution are that it can operate under any external conditions without affecting the movement of the trains, thus enabling more frequent and faster inspections and minimizing the hazards for workers and passengers without affecting the train schedule.

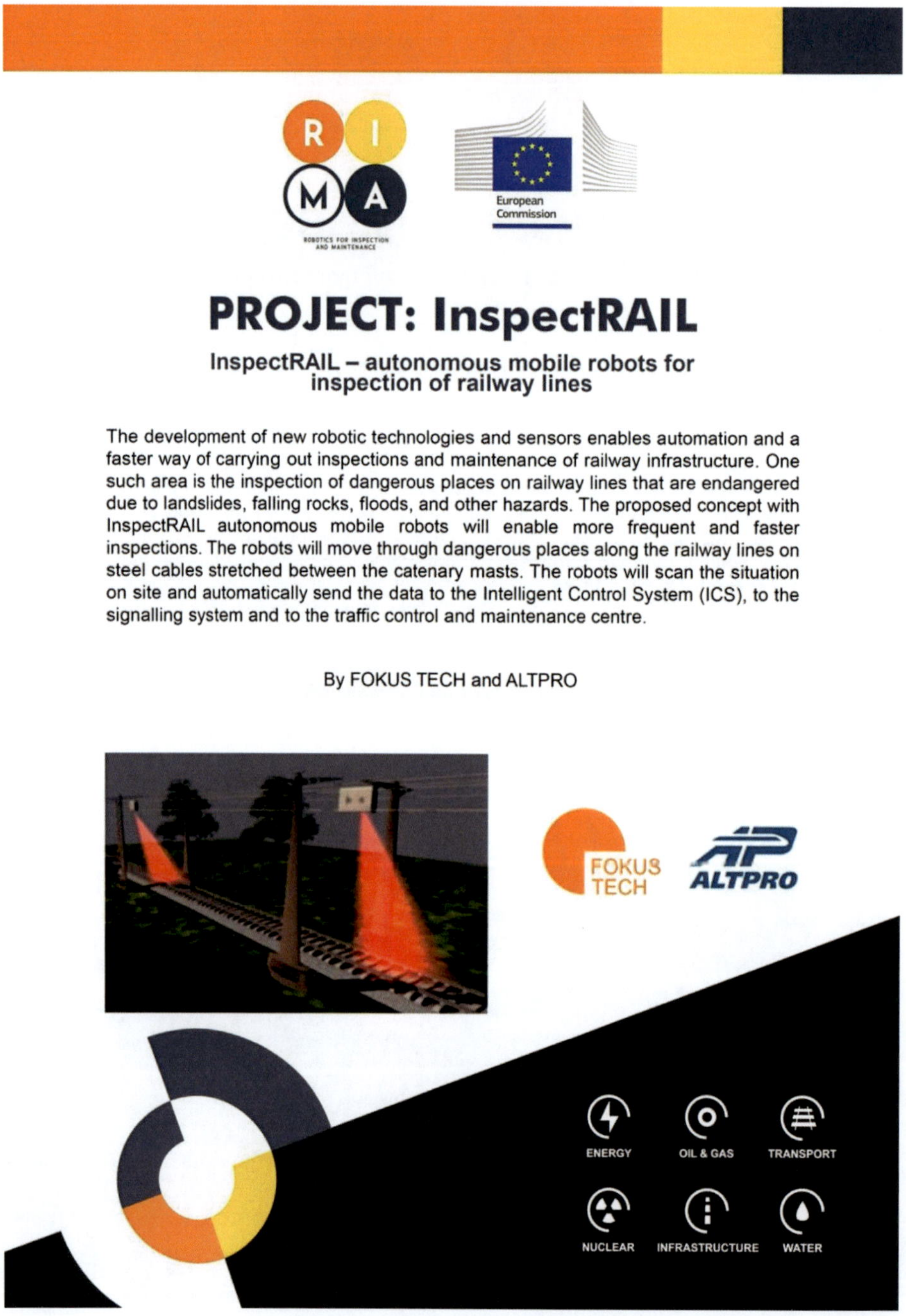

Figure 86: InspectRAIL public poster

Magic Lancer II

This application was submitted in the nuclear domain. It aimed to provide a custom mobile lightweight robotic solution made with electronics with high radiation resistance. The robot is equipped with ultrasonic, eddy current and visual testing (UT, ET, and VT) modules as well as modules for inspection of specific components. The complete system is able to withstand harsh environments, so will be capable of remote underwater non-destructive inspection of nuclear reactor pressure vessels. The benefit of this approach is the assurance of a more secure environment and reduced operator radiation exposure during different actions (e.g., sampling, dismantling, UT, ET, VT).

Figure 87: Magic Lancer II public poster

MIMIC

The MIMIC application introduced Virtual Reality (VR) to the oil & gas and chemical domain. VR enables operators to perform I&M activities remotely, controlling several remote mobile platforms at different locations through teleoperation, while the human acts physically only when needed. In this way, the efficiency, accuracy, and good planning of I&M operations is increased while the cost is reduced. Additionally, the operators have better situational awareness in the VR environment while the robot ecosystem benefits by widening its area of applications.

Figure 88: MIMIC public poster

MOVEX

The MOVEX application was submitted in the water supply and sanitation domain, to develop a system for non-destructive data collection in a sewage infrastructure and treatment plant. The solution is a modular tractor with interchangeable traction components, such as wheels and caterpillars, equipped with different sensors (e.g. sonar, infrared and other imagers) for measuring data like dimensions, pH of liquids, gas concentrations, and hydrocarbon levels. The key benefit of this solution is that the sewer does not have to be cleaned prior to inspection, unlike for the current practice of human inspection, resulting in lower cost, less downtime, and less risk of damaging the infrastructure.

Figure 89: MOVEX public poster

NYMO

The NYMO application to inspect and monitor the seabed of harbour basins and quays belongs to the transport, cargo and mobility domain. The NYMO teams developed a custom autonomous floating platform that is able to navigate itself while continuously scanning with the appropriate hardware. If abnormalities are found, such as floating obstacles or high silt levels, the system notifies the end user and provides a visualization of the relevant data through a user-friendly UI. The key benefits of the solution are: continuous reporting of the scanned area, increased efficiency, and minimizing the underwater exposure of human operators.

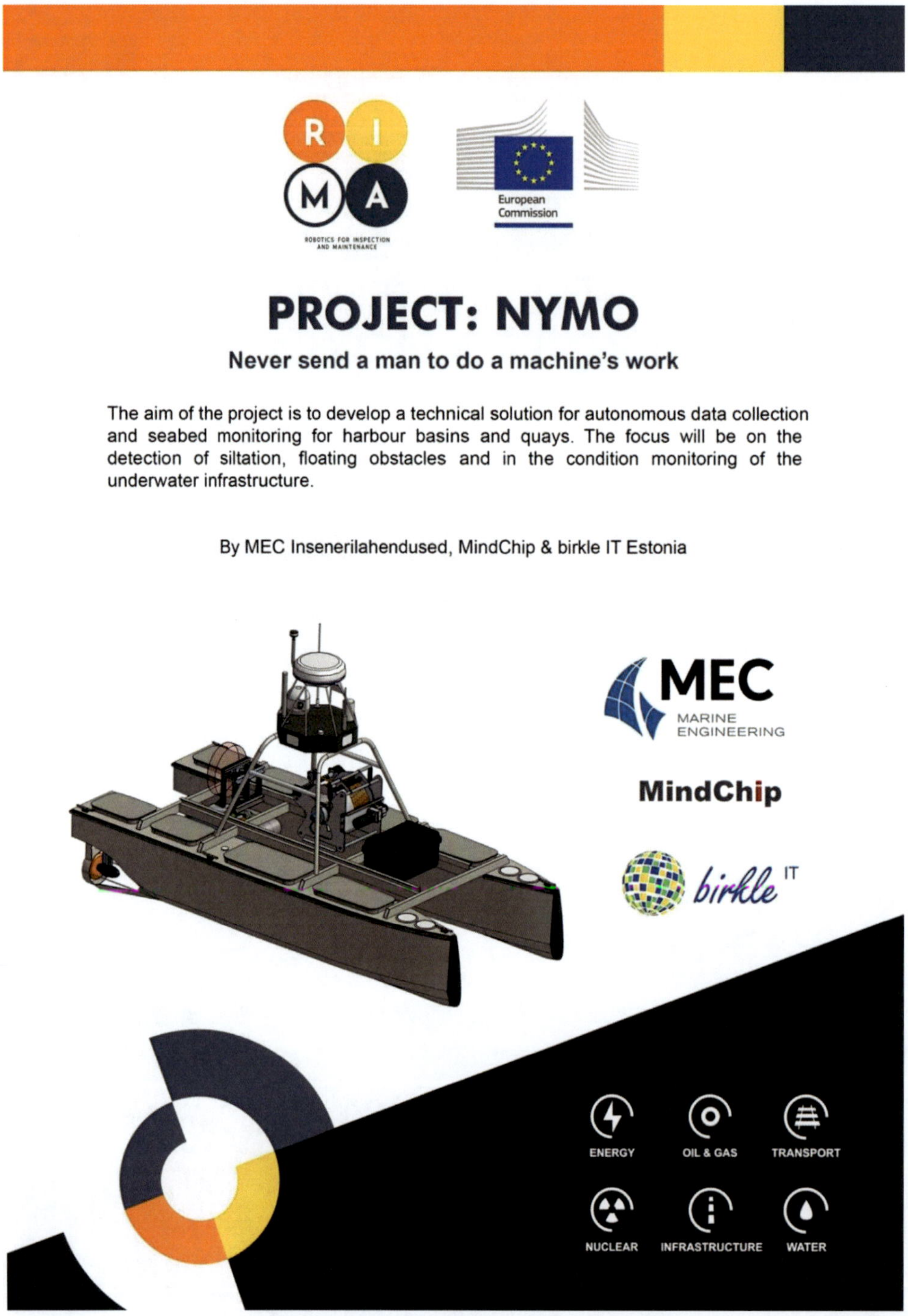

Figure 90: NYMO public poster

PIPEYE RIE

PIPEYE RIE was an application from the water supply and sanitation domain to support I&M activities in desalination plants, which are booming due to the high global demand for water. The technology involves a fully electric, battery powered ROV designed for inspecting large water pipelines from the inside, equipped with multiple sensors able to perform 3D live reconstruction during inspections. The main benefit is reducing manual inspection which require long periods in submerged and confined spaces and creates technical, operational and safety challenges. After the RIMA project, the team increased the total number of employees (also improving the female to male ratio) and signed 2 agreements and worked on 4 new ones.

Figure 91: PIPEYE RIE public poster

RAID

This application from oil & gas and chemical domain was to check petrochemical vessels for defects, assess them, and repair them if needed. A custom mobile platform able to navigate in confined areas was equipped with cameras and AI-based algorithms for automated interpretation of visual inspection and dimensioning data. The benefit of this solution is to avoid humans having to enter such areas, as stated in the safety goal of the European Chemical Industry—at least at the stage of initial detection, and also to increase the productivity of a planned inspection.

Figure 92: RAID public poster

REALIGNED

This application belongs to the nuclear domain, targeting the optimization of dismantling operations on nuclear sites. This is achieved by creating a set of local spatial anchors compliant with a universal BIM referential and a visual human-readable 360° video twin environment. This data is geotagged and timestamped through SLAM and computer vision methods and was further enhanced and updated as the project evolved. The benefit of this application is that it facilitates data recording and reporting, saving project time and cost, human resources, and the enormous data flow needed.

Figure 93: REALIGNED public poster

RoboHop

RoboHop was an application in the transport, cargo and mobility domain for performing inspections in the shipping industry. A crawler platform designed to traverse complex geometries, even if they are vertical, was equipped with sensors able to detect corrosion in the cargo tanks without the need for prior surface cleaning. The benefits of this solution are the ultra-low cost of ownership and avoiding the need to have human inspectors climbing at great heights or using rafting which wastes huge amounts of water.

Figure 94: RoboHop public poster

ROBULCLEAN

The ROBULCLEAN application in the energy generation and distribution domain cleans solar panels on photovoltaic (PV) plants. The system comprises a crane carrying a device that uses ultrasonic pressure waves to remove dirt from the PV surface. Using ultrasonic cleaning means the system uses less water and less energy than current cleaning methods. The main benefit, however, is the increased efficiency of the PV panels, since clean ones can absorb more solar power and achieve better results compared to the competition. After RIMA, the team hired 2 new employees to extend their activities.

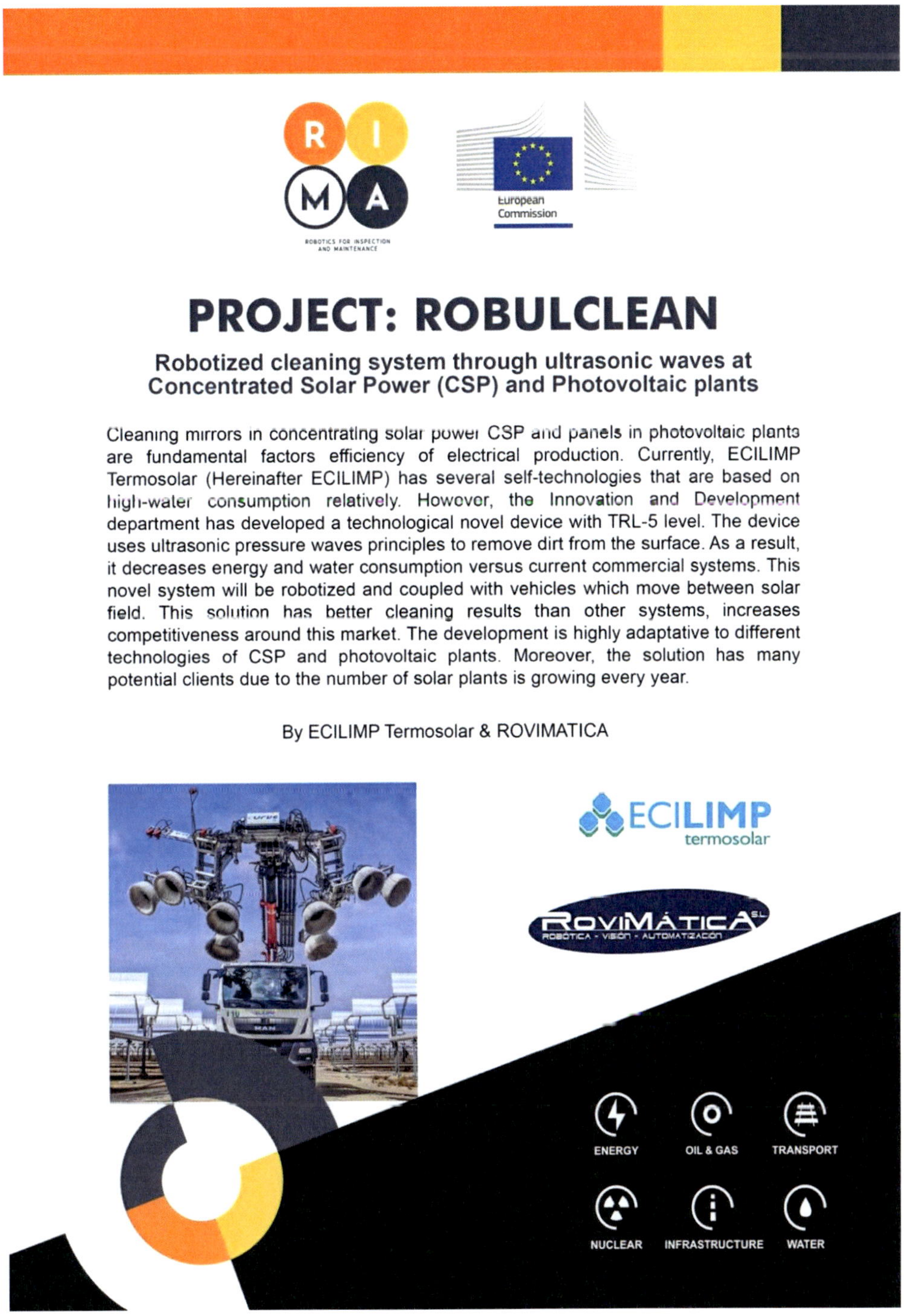

Figure 95: ROBULCLEAN public poster

RoMy

RoMy belongs to the nuclear domain and aims to perform mapping of the infrastructure to optimize processes and risk management during maintenance procedures and decommissioning operations. The technology used comprises a snake robot equipped with a 3D mapping module and a smart algorithm that fuses data from different sensors and provides a high-quality interior Geographic Information System. The key benefit is the building of reliable and complete digital twin of the infrastructure which is an important issue to ensure operations and grant the time management..

Figure 96: RoMy public poster

SBID

The SBID application in the energy generation and distribution domain, focused on the development of hydrogen-powered, fixed-wing drones that are launched from the shore to perform I&M activities on offshore windmills. Current approaches require onsite crews to do the inspection, which involves climbing on the windmills, which is unsafe. The benefit of the SBID approach is that the solution is cheaper, faster, with less environmental impact, safer for the crew, and almost independent of the weather conditions.

Figure 97: SBID public poster

SPARC I&M

Another application from the energy generation and distribution domain focused on cleaning the PV panels. SPARC I&M used a lightweight mobile platform that is positioned on the panels and can navigate autonomously. This platform is equipped with the necessary tools to perform the cleaning activities. The aim is to provide an affordable and innovative multi-robot system to reduce inspection and maintenance costs at large scale solar plants.

Figure 98: SPARC I&M public poster

SPIIDER

SPIIDER answers a challenge from the nuclear domain by focusing on the inspection of high radiation zones (HRZ) in nuclear power plants. SPIIDER used a UAV that can withstand the HRZ environment and is equipped with sensors to perform NDT measurements and an ultrasonic virtual GPS to provide precise semi-autonomous piloting, improve stability, and thus ensure a more complete inspection. The key benefits are a system that can enter areas inaccessible to humans, minimization of shutdown times, and the system's applicability to different cases.

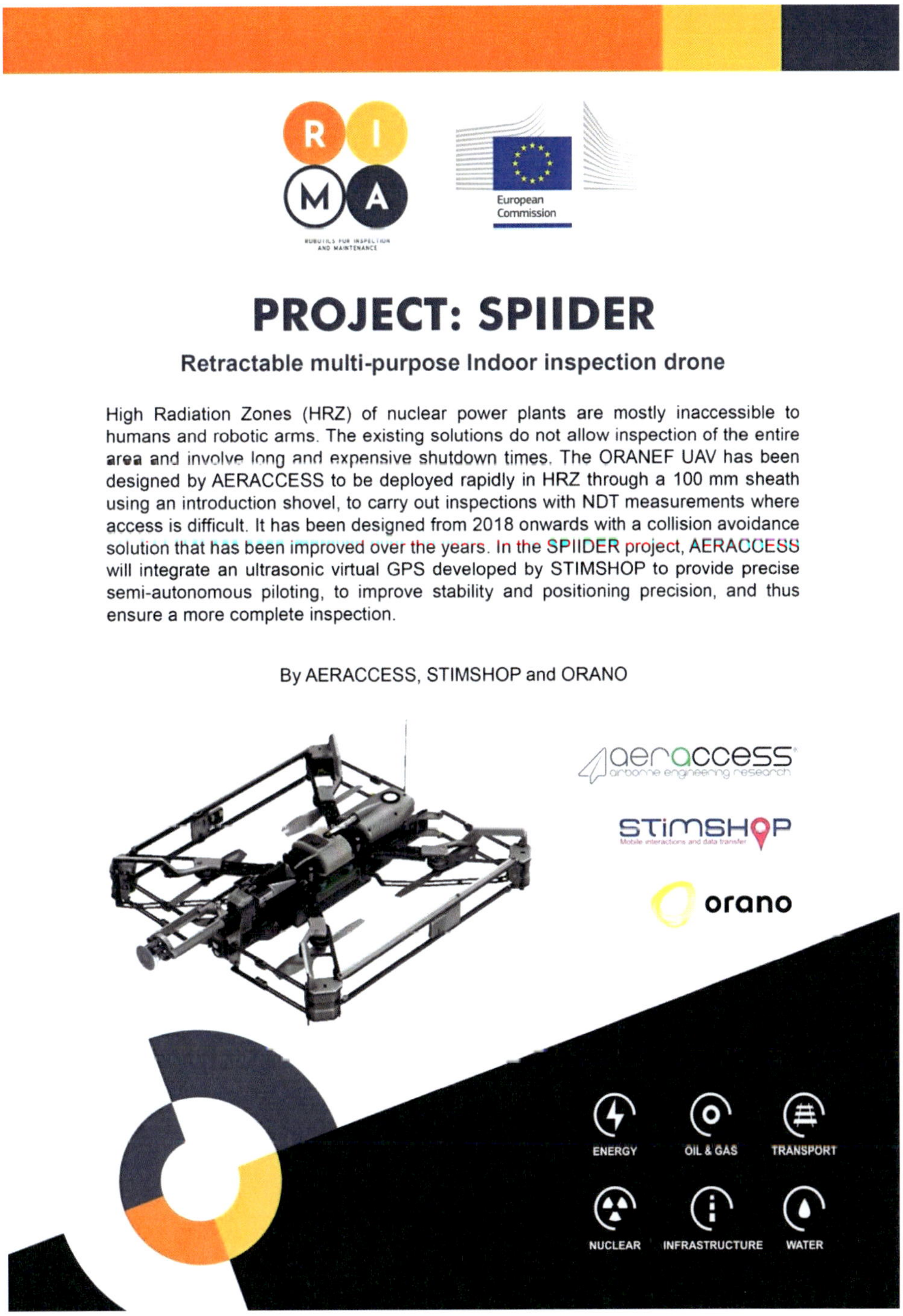

Figure 99: SPIIDER public poster

TERRAPIN

The TERRAPIN application in the water supply and sanitation domain focused on inspecting the water grid for leakages. TERRAPIN addresses this challenge by implementing a custom robotic system for trenchless water pipe inspection and ground-penetrating radar-based technology and analytics for mapping underground pipes. Defects and leakages can then be located through an automated process that has lower cost than existing solutions, avoiding laborious scanning of large areas with bulky devices.

Figure 100: TERRAPIN public poster

T-IRIS

T-IRIS was submitted in the energy generation and distribution domain. It connects the capabilities of mobile robots with the IT infrastructure used in the operation of industrial facilities. More specifically, a custom fully autonomous walking robot has been developed to be used for industrial inspection, building a digital twin of the inspected area. The benefit of the solution is that the ecosystem will profit from the establishment of best practices and software APIs that foster the integration of further solutions.

Figure 101: T-IRIS public poster

TunnelEye

The TunnelEye application from the urban and suburban transport routes and buildings domain inspected road tunnels. A multi-sensor system integrated on a vehicle was used, capable of scanning tunnel walls in high fidelity even at speeds up to 80km/h. The data captured were processed to identify defects, assess their severity and show them to human operators in the form of report. The main benefits of this approach are that there is no need to close the road while inspections take place, as there is no need for specialized inspectors to perform manual visual inspections. Furthermore, the time for the inspection is reduced by 97.5%, without even leaving the office.

Figure 102: TunnelEye public pos

Due to confidentiality issues or development ones, some esperiments do not have poster. Those are the following:

AUTODRONE

The AUTODRONE application was in the energy generation and distribution domain. The project developed an integrated module combining state-of-the-art sensors and advanced algorithms for the autonomous indoor navigation of an inspection drone in GPS-denied environments. The main advantage of the solution is that it will satisfy market demand for easy-to-perform digital inspection of indoor and confined spaces in industrial assets. It will also alleviate the need for expert pilots and enhance repeatability, while enabling widespread adoption of the technology in industry.

F3DI

F3DI was an application in the oil & gas and chemical domain aiming to demonstrate an underwater Fast-3D inspection service. The technology comprises a 3D digitalization device to create digital twins for subsea assets, using SeaVision 3D laser imaging sensor as the key technology differentiator. The benefits of the proposed solution are lower operational costs, improved maintenance strategy, improved integrity and overall production safety of the asset, and the data to enable precise planning of upgrades and life extensions. This type of solution can be extended to other domains as well, such as civil infrastructure, energy production, water distribution etc.

9.4. Lessons learned from the mentored work

General feedback and comments were collected from the applicants in the form of questionnaires as well as through discussions with the RIMA DIHs. The comments received so far could be summarized in 6 main ideas:

- The mentoring has been well organized and the availability of partners to be supportive, tolerant, and non-judgemental was well appreciated.
- We appreciated that flexibility was part of the operational and management point of view.
- SMEs encountered issues due to lack of experience or lack of contacts.
- All experiments acknowledge that DIHs have certain expertise and they were glad to use this expertise.
- Observation from real practice is that DIHs are important to SMEs.
- RIMA DIHs offered their services respecting their role and network concept.

It is worth mentioning that there were no stoppages or failures in any funded project from the 1st or the 2nd Open Call, nevertheless the RIMA network offered the flexibility to adapt and update the initial plan over the first few months due to unforeseen events, for example, difficulty in accessing certain test areas due to COVID restrictions, or delays in the delivery of parts needed to finish prototype construction so that testing could start. On average these extensions lasted 2.8 months which was addressed properly by all the mentors within the RIMA timeline. Last but not least, despite the broad range of services that the RIMA network can offer, there was a high preference by the applicants for the following services:

- Provide technical and business advice.
- Offer of infrastructure.
- Offer support on standardization, IP, legislation etc.
- Offer dissemination and network opportunities.

Last but not least, the figure below (Figure 103) summarizes several results in terms of impact, as reported by the applicants, showing the importance of the RIMA project to the applicants.

SUMMARY OF IMPACT IN NUMBERS

Almost all SMEs benefitted in an obvious manner from the RIMA technical/business and economic support, as evidenced by the numbers and information collected by the mentors. Although it is hard to track all the SMEs' progress, taking into consideration that 50 projects were funded with at least 2 SMEs each, the mentors have gathered enough information to show impact. The following list provides impact information from 10 applications from both open calls, that show in detail the post-RIMA expansion of the companies involved in different categories:

- INOWATT: A few million euros of funding received and a spin-off company created.
- Pipeye RIE: Increase in turnover by about 8%, Total number of employees and female to male ratio increased, 2 initial agreements, more details could not be disclosed, and another 4 in the pipeline.
- TANGO: Several hundred thousand euros of funding received.
- ROBULCLEAN: Hire of 2 new employees.
- PRIMUS: Creation of 1 new service, 1 new product for programming robots and a few millions of euros of funding received.
- ACIDRONES: Creation of 1 new service.
- RAHIP: Creation of 1 new product and 1 new service, total number of employees doubled.
- MAP4CAT: Increased number of personnel, gained several new customers, got a few millions of additional funding, launched 1 new product/service.
- InspectRAIL: Increased personnel numbers, gained several new customers, got a few hundred thousand euros of additional funding, launched 1 new product/service, the results of the project were presented in a paper and at an international conference.
- SPARC I&M: Have a significant increase in the number of personnel, gained several new customers, got almost a million euros of additional funding, launched 2 new product/services, and patented a new robot. The project also resulted in an IROS publication.

Overall, general information from all applications can be found below, showing the positive impact RIMA had for the SMEs, even before RIMA ended:

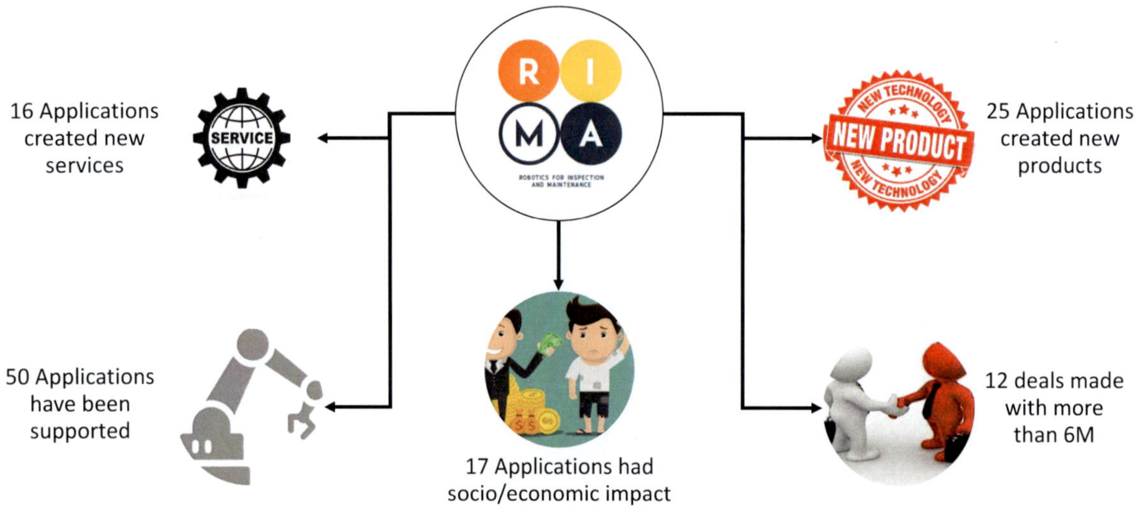

Figure 103: The reported impact of the RIMA FSTP experiments

CHAPTER 10

ROBOTIC CYBERSECURITY: LEGAL AND REGULATORY IMPLICATIONS

ANDREAS PRODROMOU AND SAVVAS LOIZOU

10.1. Introduction

Robots are complex systems which combine hardware and software, operating systems, communication processes and networks, controllers, sensors, and mechanical actuators[1] (see Figure 104). They interact with the physical world and humans. As defined in ISO 8373 2012[2], they "perform useful tasks for humans." With the ever-expanding use of robots, their interactions with humans have become more common. A significant characteristic of robotic systems is that they connect the physical and virtual worlds[3]. The highly interconnected nature of typical modern robots opens the door to cybersecurity issues and their effects on the humans interacting with them and on society at large. Michels and Walden[4] express this succinctly in their 2018 paper: "the more functions are performed across interconnected systems and devices, the more opportunities for weaknesses in those systems to arise, and the higher the risk of system failures or malicious attacks."

In developing a robot, safety is just one parameter which needs to be considered alongside development costs, market entry timing, and end-user requirements. Manufacturers also have to consider that end users place a premium on price, usability and functionality, often disregarding safety[5] until it comes back to bite them. However, the European Commission[6,7] and others[8] have shown that end users would be willing to pay for (and prioritise) high levels of security when using connected products. This prioritisation is contingent upon this security being communicated to them comprehensibly.

Using robotic and autonomous systems in everyday applications is a game changer (both within I&M[9] and beyond). Their impact will be broader than that of the computers which preceded them. Systems that can make autonomous decisions and act physically are disruptive: this capacity affects both existing and new value chains. But as robotic technology becomes more pervasive in our

---

1 D. Quarta, M. Pogliani, M. Polino, F. Maggi, A. M. Zanchettin and S. Zanero, *"An Experimental Security Analysis of an Industrial Robot Controller,"* 2017 IEEE Symposium on Security and Privacy (SP), San Jose, CA, USA, 2017, pp. 268-286. doi: 10.1109/SP.2017.20

2 ISO 8373:2012 Robots and robotic devices – vocabulary

3 E. Fosch-Villaronga, C. Millard, "Cloud robotics law and regulation: Challenges in the governance of complex and dynamic cyber–physical ecosystems," Robotics and Autonomous Systems, v. 119, 2019, pp 77-91. di: 10.1016/j.robot.2019.06.03

4 J.D. Michels, I. Walden, *"How Safe is Safe Enough? Improving Cybersecurity in Europe's Critical Infrastructure Under the NIS Directive. Improving Cybersecurity in Europe's Critical Infrastructure Under the NIS Directive,"* (December 7, 2018). Queen Mary School of Law Legal Studies Research Paper, 2018 (291)

5 G.W. Clark, M.V. Doran, T.R. Ande, *"Cybersecurity issues in robotics,"* Proceedings of the 2017 IEEE conference on cognitive and computational aspects of situation management (CogSIMA); 2017. p. 1–5

6 European Commission, Whitepaper on artificial intelligence – a European approach to excellence and trust, 2020. Available at https://ec.europa.eu/info/sites/info/files/ commission-white-paper-artificial-intelligence-feb2020_en.pdf

7 European Commission, Impact assessment on increased protection of internet-connected radio equipment and wearable radio equipment, 2020. Available at https://ec.europa.eu/docsroom/documents/40763/ attachments/2/translations/en/renditions/native

8 S.D. Johnson, J.M. Blythe, M. Manning, G.T.W. Wong, *"The impact of IoT security labelling on consumer product choice and willingness to pay,"* PLoS One, 15(1), 2020. doi:10.1371/journal.pone.0227800

9 Inspection and Maintenance

everyday lives, people are becoming less trusting of it. Users may see robots as intrusive, controlling, or dangerous to their physical well-being.

In human societies, effective laws and regulations create trust. In robotic cybersecurity, trust is not just a matter of technical functionality. The appropriate laws and regulations are necessary for robots and autonomous systems to gather widespread acceptance by users in the I&M sector: protection and regulation is a sine qua non for (a) effective robotic cybersecurity, (b) acceptance by end users, and (c) eventual mainstreaming of I&M robots.

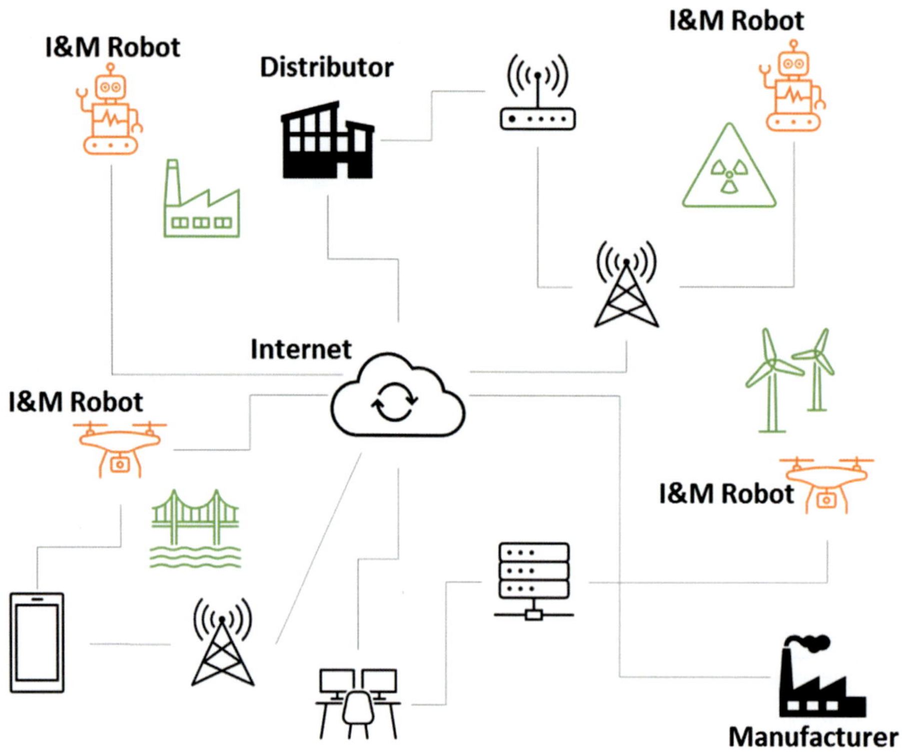

Figure 104: A high-level view of the interconnectedness of modern-day robotic systems (adapted from [1])

10.2. Cybersecurity and robots in general and I&M in particular

Cybersecurity affects all businesses at both the local and global levels. Cybersecurity is an expanding concern for robotics as robots become more mobile and connected, providing more points of entry for potential attacks. Everyone involved in robotics, including inspection and maintenance robotics, needs to consider cybersecurity and cyber-attack risks. This includes both the manufacturers of robots and their end users. The consequences of cyber attacks on robots can be significant, especially if the robots are acting on essental infrastructure. Their vulnerability can be attributed to the following robot features:

- Their reliance on communication networks to operate: the primary point of entry for cyber-attacks on all sorts of modern digital technologies (not just robots) is the vulnerabilities of communication networks. Then control of the robot can pass to the hackers.
- Their capacity for autonomous decision-making (without human oversight).
- Co-dependency can be developed between robots and humans in some instances.
- Exposure of a robot's sensing modalities to its operating environment offers an additional cyber-attack vector, allowing competitors or enemies access to live information from the site.

The co-dependency (or co-existence) between humans and robots, and the interconnected nature of robotic systems, make robots particularly susceptible to cyber-attacks. As time goes by and I&M robots are further developed, interacting directly with humans and taking over sensitive tasks (e.g., inspection and maintenance in dangerous/harsh environments or of critical infrastructures) so cybersecurity also must be developed in parallel.

Co-dependency has two levels: a personal level and a societal/industrial level. At the personal level, we refer to robots working in the same physical space as humans. If a robot in such an environment becomes subject to a cyber-attack, it could lead to the injury or death of a human (in the worst case) or some damage or loss (in the best case). It should also be noted that giving a robot a task previously performed only by a human operator creates a new cybersecurity risk where there was previously none (see Section 3.2).

Society-level cyber attacks could potentially target I&M robotic systems responsible for large-scale or critical infrastructures upon which our societies rely for their everyday activities, security and economy. Our critical infrastructures are already under attack for decades. Fortunately for us, the closed nature of most of the systems that run these infrastructures has kept us safe until now. Although introducing robotic systems for I&M and their networked/autonomous operational models brings key benefits for securing and improving I&M operations, it can present the danger of new effective cyber attacks. A hacked I&M robot could, e.g. cause damage to a power plant, a transmission network, or a pipeline. So, at the societal level, beside the physical damage and other force-majeure situations, a cyber-attack on an I&M robotic system may trigger side negative consequences such as:

- Financial damage: The end users and the robot systems manufacturers can expect financial losses due to cyber attacks targeted at robots. These losses could be particularly detrimental to SMEs and lead to bankruptcies.
- Reputational damage: Robotic systems can collect, store, and process large amounts of data. This data could be personal data but also commercially sensitive. Having data stolen will damage the reputation of a robotic company. A company's reputation could also be damaged by physical damage, human injuries, or death resulting from a cyber attack.

Cybersecurity is a hot topic globally. A patchwork of laws and regulations exist at the local, EU, and international levels. Some of these are generic, others are more focused.

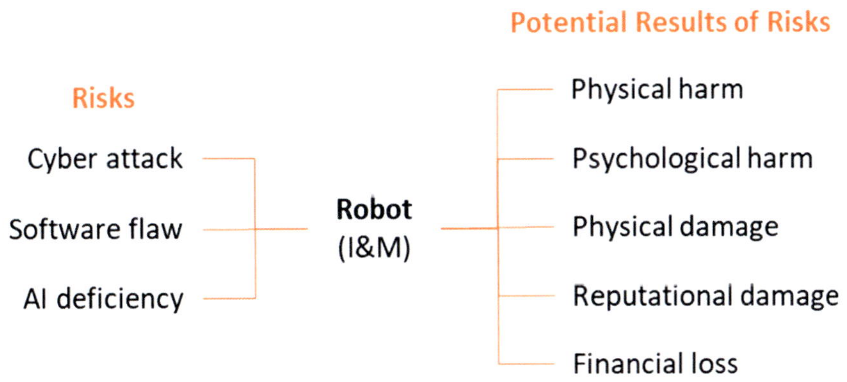

Figure 105. Various I&M robot cyber risks can result in physical and non-physical consequences

10.3. Why do we need regulation?

There are multiple answers to the question of why we need regulation in the area of robots for inspection and maintenance, each one addressing a different socioeconomic requirement:

- End users and society at large need to be provided with protection from robotic systems that may cause harm (physical, emotional, financial, other).
- Manufacturers of robots need to understand the legal framework within which they will operate. This understanding allows them to develop appropriate products without liability issues.
- To address ethical and moral issues.
- And lastly, proper regulation facilitates technological advances and can show where new business opportunities are available to be developed.

Generally speaking, lawmakers and regulators need to consider several principles succinctly described in Holder et al.[10]:

- Robotic applications have to be designed in such a way as to protect the values of human dignity, health, safety and privacy.
- Research and innovation have to be responsibly led, and the application of technologies aimed at overcoming human vulnerabilities may require legislators to review their applicability for other uses, for example, in enhancing the abilities of the already enabled; and
- Liability: Who is responsible for loss or damage caused by the acts or omissions of a robot? And should robots be treated differently from other machines?

10.4. The legal and regulatory regime in the EU

At the EU level, several instruments address cybersecurity. Of the legislation below, only the proposed Cyber Resilience Act addresses robotics directly. All the others have an indirect relation to robotics. The most relevant instruments include the following:

- *The EU Cybersecurity Act*[11]: The Cybersecurity Act introduces EU-wide rules for the cybersecurity certification of products, processes, and services. In addition, it sets a new permanent mandate for the EU Agency for Cybersecurity (ENISA). Appropriate resources have been allocated to ENISA to enable it to fulfil its goals.
- *The Directive on Security of Network and Information Systems (NIS2 Directive[12])*: The NIS2 Directive is the EU-wide legislation on cybersecurity. It provides legal measures to boost the overall level of cybersecurity in the EU.
- *The General Data Protection Regulation (GDPR[13])*: The General Data Protection Regulation (EU) provides a legal framework for keeping everyone's personal data safe by requiring companies to have robust processes in place for handling and storing personal information.
- *New telecoms rules (Electronic Communications Code[14])*: According to the rules, EU Member States have to maintain the integrity and security of public communications networks, with obligations to ensure that operators take technical and organisational measures to appropriately

10 C. Holder, V. Khurana, F. Harrison, L. Jakocs, *"Robotics and law: Key legal and regulatory implications of the robotics age (Part I of II),"* Computer Law and Security Review, v. 32, 383-402, 2016.

11 REGULATION (EU) 2019/881 OF THE EUROPEAN PARLIAMENT AND OF THE COUNCIL of 17 April 2019 on ENISA (the European Union Agency for Cybersecurity) and on information and communications technology cybersecurity certification and repealing Regulation (EU) No 526/2013 (Cybersecurity Act). Available at http://data.europa.eu/eli/reg/2019/881/oj

12 Directive (EU) 2022/2555 of the European Parliament and of the Council of 14 December 2022 on measures for a high common level of cybersecurity across the Union, amending Regulation (EU) No 910/2014 and Directive (EU) 2018/1972, and repealing Directive (EU) 2016/1148 (NIS 2 Directive). Available at http://data.europa.eu/eli/dir/2022/2555/oj

13 Regulation (EU) 2016/679 of the European Parliament and of the Council of 27 April 2016 on the protection of natural persons with regard to the processing of personal data and on the free movement of such data, and repealing Directive 95/46/EC (General Data Protection Regulation). Available at http://data.europa.eu/eli/reg/2016/679/oj

14 Directive (EU) 2018/1972 of the European Parliament and of the Council of 11 December 2018 establishing the European Electronic Communications Code Available at http://data.europa.eu/eli/dir/2018/1972/oj

manage any risks to the security of networks and services. It also requires that competent national regulatory authorities are given appropriate powers, including issuing binding instructions and ensuring compliance with them. In addition, Member States can attach conditions concerning the security of public networks against unauthorised access to the general authorisations for operators, to protect the confidentiality of communications.

- *The EU Cyber Diplomacy Toolbox*[15]: The Toolbox is a framework for a joint EU diplomatic response to malicious cyber activities that allows the EU to make full use of measures within the Common Foreign and Security Policy, including statements by the High Representative, diplomatic demarches and, if necessary, restrictive measures, in response to malicious cyber activities. The Toolbox established a sanctions regime which allows the EU to impose targeted restrictive measures to deter cyber-attacks which constitute an external threat to the EU and its Member States.
- *The Cyber Resilience Act*[16,17] *(CRA):* The Act is a law proposed by the EU to bolster the cybersecurity of digital products in the EU and address existing cybersecurity regulatory gaps. The regulation will apply to a broad framework of tangible and intangible products with digital elements. The CRA is the only EU cybersecurity regulation that explicitly addresses robotics. It will complement the NIS2 Directive, the EU Cybersecurity Act, and legislation under adoption.

A MORE IN-DEPTH LOOK AT THE CYBER RESILIENCE ACT

The CRA is a unique piece of EU-wide legislation. Once adopted, it will not only bolster the cyber-security of digital products, but non-conforming devices will be banned from the internal market. Products coming into the EU from outside must conform to the Act's requirements. The EU Parliament expects that the CRA will play for cyber-security the role the GDPR played for personal data. Many non-EU vendors are expected to find it easier to apply the CRA criteria as the default cybersecurity framework for their global operations instead of creating different products for different markets.

More critical for robotics in I&M is that, once the CRA becomes EU law, the robotic sector will have a common cybersecurity framework for all its products.

Once the European Parliament adopts the CRA, the EU member states and all interested parties will have two years to comply with its requirements. The exception to the two-year rule is the reporting obligation on manufacturers. This obligation will apply one year from the Act's entry into force.

WHAT WILL THE CYBER RESILIENCE ACT COVER?

It will apply to *"any software or hardware product and its remote data processing solutions, including software or hardware components to be placed on the market [14]."* The regulation takes a lifecycle approach - it will apply to products from design to obsolescence. The Act covers tangible digital products (e.g., connected devices) and intangible ones (e.g., the software embedded into connected devices). The CRA effectively attempts to cover all products that connect to networks (hardware and software).

---

15 Council of the EU, Press Release, 19 June 2017. Available at https://www.consilium.europa.eu/en/press/press-releases/2017/06/19/cyber-diplomacy-toolbox/

16 Proposal for a REGULATION OF THE EUROPEAN PARLIAMENT AND OF THE COUNCIL on horizontal cybersecurity requirements for products with digital elements and amending Regulation (EU) 2019/1020. Available at https://eur-lex.europa.eu/legal-content/EN/TXT/?uri=COM:2022:454:FIN

17 European Parliament Research Service, EU Legislation in Progress Briefing, “EU cyber-resilience act,” PE 739.259, December 2022.

The Act exempts specific connected devices already covered by sectoral legislation (see Article 2 of the proposed Act). Examples include the Common Rules in Civil Aviation Regulation[18] (EU 2018/1139), the Medical Devices Regulation[19] (EU 2017/745), the In Vitro Diagnostic Medical Devices Regulation[20] (EU2017/746) and the Vehicle Safety Regulation[21] (EU 2019/2144). Also exempted is software-as-a-service, unless it is an integral part of a remote data processing solution for a product with digital elements. The CRA leaves open the possibility of expanding its coverage post-enactment of the Act.

Manufacturers, importers and distributors of connected devices and services must comply with a list of cybersecurity requirements appropriate for their role and responsibilities in the supply chain. Compliance will be verified through certification, reporting, and conformity assessments.

Manufacturers must ensure that digital products comply with essential cybersecurity requirements and conformity assessment procedures before placing them on the market. In addition, they need to record technical documentation and abide by notification obligations for cybersecurity breaches.

Importers must only place digital products that comply with essential cybersecurity requirements and bear the CE marking on the market.

Distributors must verify that their digital products bear the CE marking. In addition, distributors have a duty of care to ensure that manufacturers and importers have complied with their obligations under the CRA.

The CRA has four goals:

1. Ensure that manufacturers improve the cybersecurity of products covered by the CRA throughout their whole lifecycle.
2. Create a single (and coherent) cybersecurity compliance framework for the EU.
3. Make cybersecurity practices, the properties of products, and their manufacture more transparent.
4. Provide companies and consumers with ready-to-use cyber-secure products.

The Act divides products into two categories:

1. **Default non-critical products**: This category includes products with no/low criticality levels. The EC estimates that this category covers 90 per cent of connected devices. Examples include

18 Regulation (EU) 2018/1139 of the European Parliament and of the Council of 4 July 2018 on common rules in the field of civil aviation and establishing a European Union Aviation Safety Agency, and amending Regulations (EC) No 2111/2005, (EC) No 1008/2008, (EU) No 996/2010, (EU) No 376/2014 and Directives 2014/30/EU and 2014/53/EU of the European Parliament and of the Council, and repealing Regulations (EC) No 552/2. Available at http://data.europa.eu/eli/reg/2018/1139/oj

19 Regulation (EU) 2017/745 of the European Parliament and of the Council of 5 April 2017 on medical devices, amending Directive 2001/83/EC, Regulation (EC) No 178/2002 and Regulation (EC) No 1223/2009 and repealing Council Directives 90/385/EEC and 93/42/EEC. Available at https://eur-lex.europa.eu/legal-content/EN/TXT/?uri=CELEX%3A32017R0745

20 Regulation (EU) 2017/746 of the European Parliament and of the Council of 5 April 2017 on in vitro diagnostic medical devices and repealing Directive 98/79/EC and Commission Decision 2010/227/EU. Available at http://data.europa.eu/eli/reg/2017/746/oj

21 Regulation (EU) 2019/2144 of the European Parliament and of the Council of 27 November 2019 on type-approval requirements for motor vehicles and their trailers, and systems, components and separate technical units intended for such vehicles, as regards their general safety and the protection of vehicle occupants and vulnerable road users, amending Regulation (EU) 2018/858 of the European Parliament and of the Council and repealing Regulations (EC) No 78/2009, (EC) No 79/2009 and (EC) No 661/2009 of the European Parliament and of the Council and Commission Regulations (EC) No 631/2009, (EU) No 406/2010, (EU) No 672/2010, (EU) No 1003/2010, (EU) No 1005/2010, (EU) No 1008/2010, (EU) No 1009/2010, (EU) No 19/2011, (EU) No 109/2011, (EU) No 458/2011, (EU) No 65/2012, (EU) No 130/2012, (EU) No 347/2012, (EU) No 351/2012, (EU) No 1230/2012 and (EU) 2015/166. Available at https://eur-lex.europa.eu/eli/reg/2019/2144/oj

video games/consoles or smart home assistants. The companies responsible for these products must self-assess their vulnerabilities for improvement.

2. **Critical products:** These are subdivided into two categories. Class I are the lower-risk products, and Class II are the higher-risk ones.

Annex III of the CRA contains a list of Class I and Class II products. This list currently contains the following:

Class I:

1. Identity management systems software and privileged access management software
2. Stand-alone and embedded browsers
3. Password managers
4. Software that searches for, removes, or quarantines malicious software
5. Products with digital elements with the function of a virtual private network (VPN)
6. Network management systems
7. Network configuration management tools
8. Network traffic monitoring systems
9. Management of network resources
10. Security information and event management (SIEM) systems
11. Update/patch management, including boot managers
12. Application configuration management systems
13. Remote access/sharing software
14. Mobile device management software
15. Physical network interfaces
16. Operating systems not covered by class II
17. Firewalls, intrusion detection and/or prevention systems not covered by class II
18. Routers, modems intended for the connection to the internet, and switches, not covered by class II
19. Microprocessors not covered by class II
20. Microcontrollers
21. Application-specific integrated circuits (ASIC) and field-programmable gate arrays (FPGA) intended for the use by essential entities of the type referred to in Annex I to the NIS2 Directive
22. Industrial Automation & Control Systems (IACS) not covered by class II, such as programmable logic controllers (PLC), distributed control systems (DCS), computerised numeric controllers for machine tools (CNC) and supervisory control and data acquisition systems (SCADA)
23. Industrial Internet of Things devices not covered by class II

Class II:

1. Operating systems for servers, desktops, and mobile devices
2. Hypervisors and container runtime systems that support virtualised execution of operating systems and similar environments
3. Public key infrastructure and digital certificate issuers
4. Firewalls, intrusion detection and/or prevention systems intended for industrial use
5. General purpose microprocessors
6. Microprocessors intended for integration in programmable logic controllers and secure elements
7. Routers, modems intended for the connection to the internet, and switches, intended for industrial use
8. Secure elements
9. Hardware security modules (HSMs)

10. Secure cryptoprocessors
11. Smartcards, smartcard readers and tokens
12. Industrial automation & control systems (IACS) intended for the use by essential entities of the type referred to in Annex I to the NIS2 Directive, such as programmable logic controllers (PLC), distributed control systems (DCS), computerised numeric controllers for machine tools (CNC) and supervisory control and data acquisition systems (SCADA)
13. Industrial Internet of Things devices intended for the use by essential entities of the type referred to in Annex I to the NIS2 Directive
14. Robot sensing and actuator components and robot controllers
15. Smart meters.

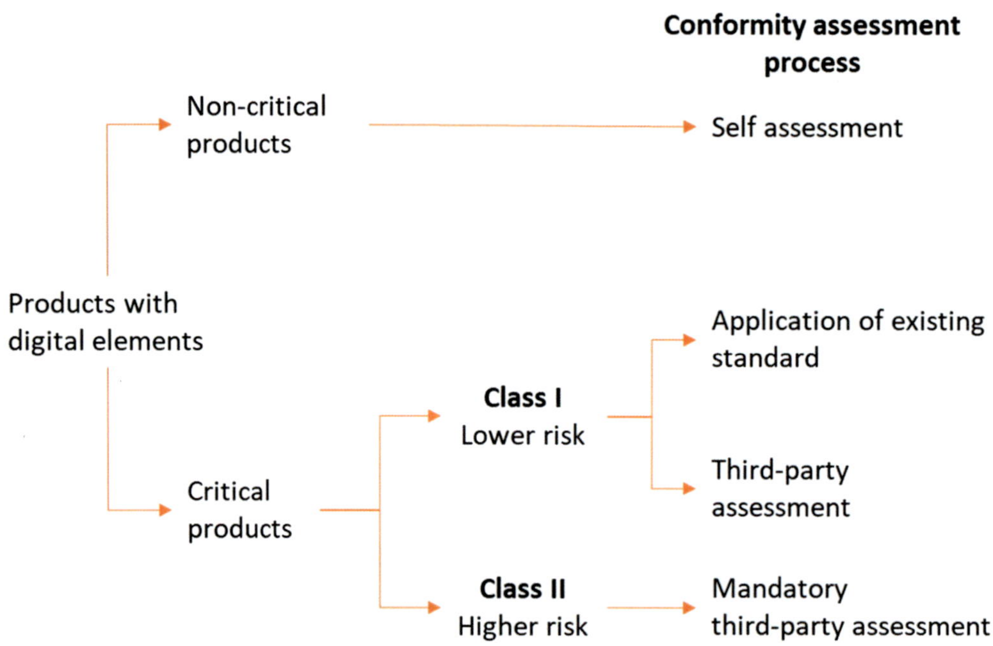

Figure 106. The conformity assessment process of the Cyber Resilience Act

WHAT ARE THE MAIN PROVISIONS OF THE CYBER RESILIENCE ACT?

The primary provisions of the CRA are presented in the following subsections.

Security by design and by default

Manufacturers must consider cybersecurity from the design and development phase of the digital product by using secure-by-default configurations and avoiding known exploitable vulnerabilities. The annexes[22] of the proposed CRA specifies the following:

1. The information manufacturers must make available to users.
2. The conformity assessment procedures digital products must go through.
3. The technical documentation that the manufacturers must provide.
4. The vulnerability handling requirements manufacturers must follow to ensure the cybersecurity of their digital products.

22 COM(2022) 454, "Annexes to the proposal for a regulation of the European Parliament and of the Council on horizontal cybersecurity requirements for products with digital elements and amending Regulation (EU) 2019/1020," Brussels 15.09.2022. Available at EUR-Lex - 52022PC0454 - EN - EUR-Lex (europa.eu).

Essential cybersecurity and vulnerability handling and reporting requirements

The CRA splits the cybersecurity obligations for manufacturers into two: (a) security requirements relating to the properties of digital products and (b) vulnerability handling requirements.

Annex I lists the cybersecurity requirements manufacturers must adhere to. Among the obligations included in the list are the following:

1. Design, develop, and manufacture digital products in such a way that limits their attack surface and reduces the impact of any incident, based on a risk assessment.
2. Deliver digital products without known exploitable vulnerabilities.
3. Protect the confidentiality and integrity of data stored, transmitted or processed.
4. Process only data (personal or other) that are strictly necessary to the functioning of the digital product (the 'data minimisation' principle).

Following the market introduction of the product, manufacturers must:

1. Put in motion a regular test and review process for the security of their digital products.
2. Keep a record of identified vulnerabilities.
3. Fix vulnerabilities through the provision of free updates.

The above must be available for either (a) the expected lifetime of the digital product or (b) for a period of five years (whichever is shorter).

Digital product manufacturers will also have to report to ENISA (the European Union Agency for Cybersecurity) all actively exploited vulnerabilities and security incidents within 24 hours of becoming aware of them.

Conformity assessment and compliance

The criticality of the digital product decides which conformity assessment procedure must be used to demonstrate compliance with the requirements set forth by the CRA (see Figure 106):

- For non-critical products, manufacturers are responsible for declaring that their products satisfy the essential security.
- For critical products, manufacturers must apply harmonised security standards (e.g. the EU cybersecurity certification scheme) or provide a third-party conformity assessment (the Member States will designate these third parties).

A digital product demonstrating compliance with (a) the security requirements and (b) the conformity assessment process will obtain an EU declaration of conformity. This declaration will be valid in all EU Member States and bear the CE marking (per the general principles of Regulation (EC) 765/2008[23]).

Fines

The CRA dictates that the Member States must appoint market surveillance authorities responsible for enforcing the proposed CRA obligations. In cases of non-compliance, companies will be fined based on the nature of the economic operator and the type of infringement.

---

23 Regulation (EC) No 765/2008 of the European Parliament and of the Council of 9 July 2008 setting out the requirements for accreditation and market surveillance relating to the marketing of products and repealing Regulation (EEC) No 339/93. Available at http://data.europa.eu/eli/reg/2008/765/oj

- Manufacturers of digital products could face fines of up to €15 million or 2.5 % of their total annual turnover worldwide, whichever is higher, for non-compliance with the security requirements listed in Annex I.
- Manufacturers, importers, or distributors could face a fine of €10 million or 2 % of their total annual turnover worldwide, whichever is higher, for non-compliance with any other obligation laid down in the CRA.

The interaction between the conformity assessment procedure of the CRA and existing or upcoming cybersecurity legislation

The proposed CRA and its conformity assessment interlock with other existing or proposed cybersecurity legislation. The proposal aims to harmonise the EU regulatory landscape by introducing cybersecurity requirements for products with digital elements. These new cybersecurity requirements do not overlap with requirements from the following existing and upcoming legislation pieces.

Existing legislation

The CRA complements the baseline EU cybersecurity framework, namely the NIS2 Directive and the EU Cybersecurity Act. The *NIS2 Directive* sets up cybersecurity requirements and incident reporting obligations for essential entities to increase their resilience (e.g., essential entities must demonstrate how they have assessed the security level of the ICT products and services). The enhanced and certified level of cybersecurity of products with digital elements (resulting from the CRA) would facilitate compliance by the entities covered by the NIS2 Directive. This would strengthen the security of the entire supply chain.

The EU Cybersecurity Act allows the development of voluntary certification schemes. Each scheme references relevant standards, technical specifications and other cybersecurity requirements defined in the certification approach. Digital products respecting such voluntary cybersecurity certification schemes would be assumed to comply with the proposed CRA's conformity assessment.

The CRA also applies to radio equipment within the scope of the RED delegated regulation[24] (EU 2022/30). The Act aligns with the requirements of the *RED delegated regulation* imposing high-level standards on internet-connected wireless and wearable radio equipment manufacturers. To avoid a regulatory overlap, the Commission will repeal the RED delegated regulation aspects concerning specific radio equipment that is also covered by the draft CRA regulation once it enters into force.

Legislation in the process of being adopted

The CRA conformity assessment will consider the provisions of the proposed Artificial Intelligence Act[25] and the (proposed) Regulation on Machinery Products[26]. The conformity assessment procedure specified by the CRA will apply to devices classified as high-risk AI systems. This is so such devices can demonstrate compliance with the security requirements of the proposed AI Act (exceptions exist). Digital products also covered in the proposed Regulation on Machinery Products

---

24 Commission Delegated Regulation (EU) 2022/30 of 29 October 2021 supplementing Directive 2014/53/EU of the European Parliament and of the Council with regard to the application of the essential requirements referred to in Article 3(3), points (d), (e) and (f), of that Directive. Available at http://data.europa.eu/eli/reg_del/2022/30/oj

25 Proposal for a Regulation of the European Parliament and of the Council Laying Down Harmonised Rules on Artificial Intelligence (Artificial Intelligence Act) and Amending Certain Union Legislative Acts. Available at https://eur-lex.europa.eu/legal-content/EN/TXT/?uri=celex%3A52021PC0206

26 Proposal for a Regulation of the European Parliament and of the Council on Machinery Products. Available at https://ec.europa.eu/docsroom/documents/45508

(for which a conformity assessment is required) will be considered to conform with the proposed CRA if the health and safety requirements of the machinery regulation are met.

10.5. Cybersecurity and the interplay of AI and robotics

The interplay of AI and robot cybersecurity is a big unknown. AI is a strategic technology for all data-driven organisations and systems, including robotic systems. Making sure that AI is secure is vital for businesses and robot operations. AI opens unique cyber-attack pathways that are not yet fully understood or addressed. Interestingly, there has been a documented increase[27] in private investment in AI cybersecurity over the past few years (2017-2022).

AI systems have two critical components: Big Data and a data model. Large amounts of data are needed to train learning systems to provide accurate answers. The data model results from training an algorithm with data, a good model requires sufficient and sufficiently varied data. A company must secure the data and model to protect AI products and associated robots. What does this mean in practice?

Data poisoning: AI systems rely on large data sets. All users of AI and robots need to ensure that their data sets are not compromised; otherwise, the AI system may be manipulated to provide incorrect results.

On- and offline system manipulation and retraining: An AI system may be manipulated by providing malicious input designed to make it provide the wrong output. Such manipulation is probably the most severe cybersecurity issue facing an AI system. It allows an attacker to exploit the AI system over a network without gaining access to the underlying data sets or code. Early public versions of generative AI systems, such as those using ChatGPT[28] (used in Bing[29]) and LaMDA[30] (used in Bard[31]), demonstrated manipulative behaviour from users trying to test the limits of the systems. Bing was, e.g. tricked into revealing the rules governing its responses[32].

Learning attack: Most AI systems use a pre-trained learning model to fulfil the tasks they were developed for. Transfer/Injection learning[33] is the application of knowledge gained from completing one task to help solve a different but related problem. An attacker can use knowledge of the learning model to attack the system and change its behaviour.

10.6. Recommendations

Interconnected robots - including I&M robots – operating in or outside of factories and/or infrastructures (including critical ones) are a relatively new thing. Most current laws and regulations regarding safety and security were designed for individual robotic systems operating in isolated industrial environments. Safety and cybersecurity are usually addressed in separate and disparate

27 N. Maslej, L. Fattorini, E. Brynjolfsson, J. Etchemendy, K. Ligett, T. Lyons, J. Manyika, H. Ngo, J. C. Niebles, V. Parli, Y. Shoham, R. Wald, J. Clark, and R. Perrault, *"The AI Index 2023 Annual Report,"* AI Index Steering Committee, Institute for Human-Centered AI, Stanford University, Stanford, CA, April 2023.

28 Chat Generative Pre-trained Transformer developed by the OpenAI corporation.

29 www.bing.com.

30 Language Model for Dialogue Application developed by Google.

31 bard.google.com.

32 L. Tung, *"Microsoft's ChatGPT-powered Bing argues with users, reveals confidential information,"* ZDNet, Feb. 2023. Available at www.zdnet.com.

33 The term transfer learning is used in conjunction with machine learning. In deep learning, the term injection learning is used. Regardless of the terminology, learning attacks are one of the core cybersecurity issues faced by AI systems.

pieces of legislation. To make robots safe, both their digital and physical parts need safeguards[34]. Why? Because they have an immediate material effect on the environment in which they operate.

Empirically, the majority of robotics practitioners agree that most robots in current use are vulnerable. Most approaches to mitigate potential attacks focus primarily on the perimeter. Experts disagree on whether this castle-and-moat approach is effective for connected robots. The reality is that we have yet to see a concerted cyber attack on robots.

Robotic systems are complex (and are getting even more complicated). This complexity makes robotic cybersecurity problematic, leading to broad attack surfaces and many attack vectors. Most manufacturers are failing to address these.

Some robot manufacturers are showing increasing concern for cybersecurity, but others demonstrate a lack of concern and investment in cybersecurity (robot manufacturers' marketing messages, openness, and community do not protect from cyber-attacks).

The EU has recognised that the current legal framework does not adequately address cybersecurity and physical product/human safety issues associated with modern connected robots. To address this weak link, the EU is in the process of adopting the Cyber Resilience Act. Its practical implementation is at least two years down the road.

The maturity and effectiveness of current defence mechanisms are questionable. Considering the above, the main overall recommendation for robot cybersecurity is to implement the adage that the best defence is a good offence. Therefore, based on the analysis conducted in the RIMA project, a set of detailed recommendations have been formulated for end users of robotic systems, as described in the next subsections.

Implement a Zero Trust approach towards robot cybersecurity

Zero Trust is a cybersecurity model that assumes all users, devices, and applications are untrustworthy until proven otherwise. It is based on the principle of "never trust, always verify," i.e., no user or device should be trusted by default: it is assumed that the system will be breached, and security is designed as if there is no perimeter[35]. A Zero Trust security framework requires all users, within and outside the organisation's network, to be authenticated, authorised, and continuously validated before being granted access to resources (e.g. robots, applications, data). Zero Trust architecture reduces risk by using robust identity verification, validating the compliance of a device/resource before granting access and ensuring “least privilege” access, where people and devices gain access to explicitly authorised resources only.

A key objective of Zero Trust is to reduce the "blast radius." The term is used to designate the impact that a security breach in one resource would have on the overall system. The aim is that no security breach should bring the whole system to its knees, i.e. the system should still be able to operate in some manner after an attack.

There is no globally accepted definition for Zero Trust. Nevertheless, the US NIST[36] Special Publication 800-207[37] seems to be a popular basis for many Zero Trust architecture formulations. NIST SP 800-207 sets seven key technology-agnostic tenets for a Zero Trust architecture:

1. All data sources and computing services are considered resources.

---

34 S. Morante, J.G. Victores, C. Balaguer, *“Cryptobotics: why robots need cyber safety,”* Frontiers in Robotics and AI, 2:23, 2015. doi: 10.3389/frobt.2015.00023.
35 Traditional cybersecurity relies upon perimeter defences – firewalls. Firewalls control traffic in and out of a network. This was based on the “trust, but verify” approach.
36 National Institute of Standards and Technology, www.nist.gov.
37 S. Rose, O. Borchert, S. Mitchell, S. Connelly, *“Zero Trust Architecture,”* Special Publication (NIST SP) 800-207, National Institute of Standards and Technology, 2020. doi: 10.6028/NIST.SP.800-207.

2. All communication is secured regardless of network location.
3. Access to individual resources is granted on a per-session basis.
4. Access to resources is determined by dynamic policy – including the observable state of client identity, application/service, and the requesting asset – and may include other behavioural and environmental attributes.
5. The enterprise monitors and measures the integrity and security posture of all owned and associated assets.
6. All resource authentication and authorisation are dynamic and strictly enforced before access is allowed.
7. The enterprise collects as much information as possible about the current state of assets, network infrastructure and communications and uses it to improve its security posture.

Even though different definitions of what constitutes Zero Trust exist, they are more or less similar. For example, the UK's National Cyber Security Centre[38] (NCSC) has a similar yet different approach to Zero Trust. The Zero Trust principles[39] of the NCSC are the following:

1. A single strong source of user identity
2. User authentication
3. Machine authentication
4. Additional context, such as policy compliance and device health
5. Authorisation policies to access an application
6. Access control policies within an application

Europe has lagged behind in actually implementing Zero Trust, focusing more on theoretical analyses. Fortunately, things seem to be changing[40], with organisations turning their attention to how to implement Zero Trust. Forrester's 2022 Security Survey shows that over 66% of European enterprises have begun developing a Zero Trust strategy[41].

Set up periodic cybersecurity assessments

Cybersecurity assessments should be carried out at least once a year, covering all users, devices, and applications. Ideally, reviews should occur whenever device or application changes are implemented. There are three primary types of evaluations which should be carried out (external organisations should be used in addition to a company's internal specialists):

1. **Security audits:** Such audits evaluate an organisation's security controls against established standards.
2. **Vulnerability assessments:** These assessments are designed to identify vulnerabilities in devices and applications.
3. **Penetration testing**: Penetration tests simulate real-world attacks on devices and applications. They should be conducted after security audits, vulnerability assessments, and improvements are implemented. It should be noted that penetration testing must have both virtual and physical components, as intruders may attempt to physically access a target device, such as a robot instead of accessing it over a network.

38 www.ncsc.gov.uk.
39 *"Network Architectures,"* National Centre for Cyber Security. Available at https://www.ncsc.gov.uk/collection/device-security-guidance/infrastructure/network-architectures.
40 T. Olufon, *"Zero Trust Comes Into The Mainstream In Europe,"* Forrester Trends Report, March 2023. Available at https://www.forrester.com/report/Zero Trust-comes-into-the-mainstream-in-europe/RES178958
41 T. Olufon, *"Zero Trust Comes Into The Mainstream In Europe,"* Forrester Blogs, March 2023. Available at https://www.forrester.com/blogs/Zero Trust-comes-into-the-mainstream-in-europe/

Protect AI systems

Several steps can be taken to protect AI systems from malicious actors. These can be summed up as follows:

1. Protect training data by controlling access so hackers cannot manipulate it.
2. Maintain data ownership records to limit the possibility of system retraining.
3. Control the input of unverified data to stop system manipulation.
4. Monitor the behaviour of the AI system for unexpected/suspicious machine learning behaviour, which could indicate a transfer learning attack.
5. Put policies in place to control data and function extraction.

An organisation must create an extremely secure computing environment to secure its AI system and components. A robust end-to-end data protection system must be at the heart of such a system.

SECTION D

SUSTAINABILITY, LESSONS LEARNED, PERSPECTIVES, CONCLUSIONS

CHAPTER 11

NETWORK SUSTAINABILITY, EXTENSION OF THE NETWORK, THE ALLIANCE

MAARIT SANDELIN

11.1. Requirements for I&M Digital Innovation Hubs

The concept of DIHs is that they are dynamic—constantly adapting to the changing needs of the regions where they are located and to the shifting needs and agendas at national and European level. Thus, it is difficult to list a concrete set of requirements for being a DIH. Up until now, the criteria of the Smart Specialisation Platform (S3 platform) for DIHs have been widely accepted. According to this platform1, an organisation must fulfil the following 4 requirements to be characterized as a DIH, and currently being evolved to EDIH:

1. Be part of a regional, national, or European policy initiative to digitise the industry.
2. Be a not-for-profit one-stop shop that supports companies.
3. Have a physical presence in the region and present an up-to-date website clearly explaining the DIH's activities and services provide related to the digital transformation of SMEs/midcaps or industrial sectors – especially those that are slow in taking up digital technologies.
4. Have at least 3 examples of how the DIH has helped a company with their digital transformation, referring to publicly available information, identifying for each: Client profile, client need, and the solution provided to meet that need.

It is widely recognised that the needs of the regions where the DIHs are located (or will be established) should be a high priority when deciding what it requires to be able to call oneself a DIH2.

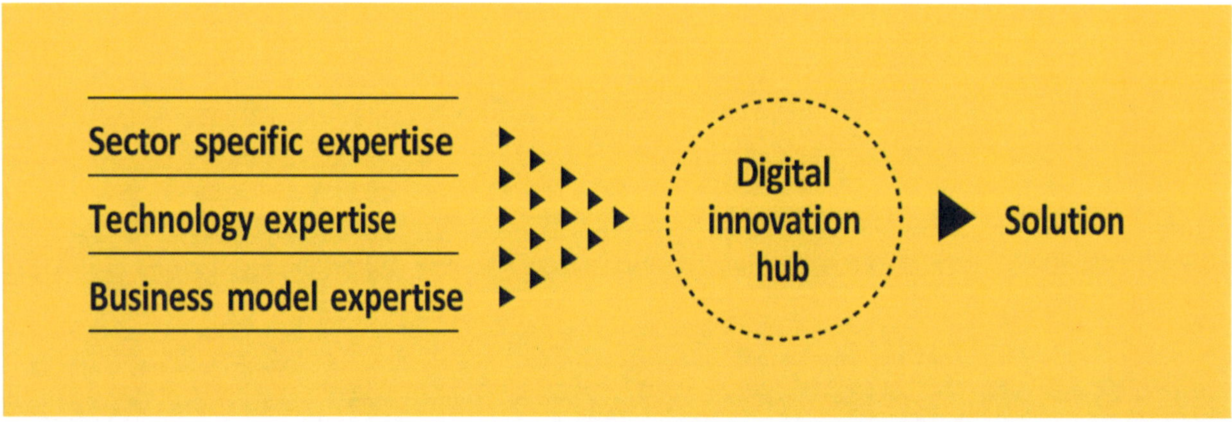

Figure 107: The DIH model (source: Roundtable on DEI Working Group 1 on DIH)

A debate on DIHs has increasingly recognised that it is difficult for a single organization to fulfil all the requirements to be characterized as a DIH. Therefore, there is a strengthened emphasis on DIHs needing to be ecosystems of various partners. DIHs that are single organisations or groups of

---

1 S3 Platform for DIHs, https://european-digital-innovation-hubs.ec.europa.eu/edih-catalogue

2 https://ec.europa.eu/digital-single-market/en/news/digital-innovation-hubs-helping-companies-and-public-administrations-make-most-digital

organisations are both characterised in the conceptual operating model of DIHs presented in Figure 107.

Obviously besides supporting companies, one could say that this model includes the most basic requirements towards being or establishing a DIH: the necessity to combine sector expertise, technology expertise and business expertise in a hub with access to relevant (digital) infrastructure, to provide the best solutions, training and advice to businesses in a given region and/or sector. For DIHs operating in the field of robotics for I&M, we suggest that they would need to combine sector-specific operational expertise (e.g. through partners or stakeholders using the DIH) with technical expertise on robotics (e.g. research expertise) and business expertise (e.g. on how the technology could help transform and/or impact a business model). This, combined with access to infrastructure to enable testing and validation of suggested solutions and/or technologies, gives a great foundation for DIH service provision. Examples of services a DIH can provide based on this combination of expertise and infrastructure are as follows:

- An SME is developing a UAV-based system for inspection. The SME is competent in UAV development but lacks competence in sense&avoid technology. The DIH supports the SME in developing a sense&avoid system for the drone, by e.g., providing cameras and other types of sensing systems or licenses for vision and decision-making algorithms.
- An SME has developed a great prototype but lacks proper test facilities to validate and prove its product. A DIH supports the SME by connecting it with an end user who has relevant test facilities and has an interest in using the product.
- An SME has a very good idea for a product, but lacks suitable partners, knowledge of the market and a strategy for commercialization. The DIH supports the SME with identifying these aspects and developing a business plan and a strategy for commercialization with an IPR strategy.

The concept of a DIH as a regional hub has become quite well established in Europe over the last few years, however there is currently a growing focus on the pan-European collaboration between the hubs. This is very clear in the concept of European Digital Innovation Hubs (EDIH), which was implemented in the Digital Europe Programme3. The Digital Europe "European Digital Innovation Hubs" call (DIGITAL-2021-EDIH-01) aimed to ensure that there is a hub in every region, and then connect these hubs on a European level to create a DIH network of scale, to be in a better position to compete with larger markets such as U.S. and Asia.

LESSONS LEARNED FROM COMMUNITY ENGAGEMENT

To engage and attract companies from all over Europe and connect them to RIMA project, an engagement strategy was formed and articulated to the entire RIMA network to be used by all RIMA partners. This strategy included communication activities such as info sessions and different events held regionally all over Europe to present the RIMA project and services it provided. A visual identity for the promotional material was created and shared with all RIMA consortium partners. Multiple media activities from social media posts to newspaper and technical articles have also been published during the project.

Communication activities were executed in close cooperation with all RIMA consortium partners, who followed the communication strategy in their own regions.

The RIMA project involves various actors: consortium partners all over Europe, start-ups, SMEs, slightly bigger companies, funders, etc. Communication with these different target groups has required different types of communication activities and it required considerable effort to reach all the target groups. Even getting all the 13 DIHs informed and motivated to take part in the communication activities on top of their own responsibilities required quite a lot of effort. New kinds

3 https://digital-strategy.ec.europa.eu/en/activities/digital-programme

of communication and motivation activities were required during the COVID pandemic because we were not able to meet in person. The communication strategy needed to be quickly modified into a more virtual form.

Despite the challenges faced, the RIMA project has been a great vantage point from which to see the current state of robotics for I&M in Europe through the eyes of DIHs, research institutes, start-ups, SMEs and slightly bigger companies. The communication activities in this type of big EU-project requires a lot of cooperation, open and regular communication, common motivation and a will to try one's best to make a difference in order for the communication to be successful—on one hand to avoid doing overlapping activities but on the other hand to do all activities that were needed.

11.2. RIMA network sustainability

11.2.1. Framing the sustainability

Post-project sustainability has been a goal from the beginning of the project, supported by all partners. It has been further motivated by the results: the successes and achievements of the project and from the options offered to the partners to further collaborate and cooperate in I&M robot applications. Maintaining the RIMA network is of interest for all RIMA partners as the network has shown the potential of

- Gathering organizations involved in research and innovation in I&M robotics
- Sharing practices that support innovation with a focus on SMEs
- Sharing information about opportunities for funding of RDI projects
- Sharing information about progress in the state of the art
- Sharing information about the challenges existing in I&M robotics

Analysis of the best way to make the RIMA network sustainable started very early in the project. Different solutions where considered, bearing in mind that network sustainability needed to be:

- Achievable before the end of the project.
- Inclusive for all partners
- Of convincing interest to partners, to the EC and to potential partners
- Complementary to existing organizations

This led RIMA to exclude solutions that would take too much time or be rejected by some partners. The consortium's original suggested method was the integration of RIMA Network into the existing not-for-profit SPRINT Robotics Collaborative, and using their already existing platform to continue RIMA in a sustainable way. In the end this option was excluded due to not achieving consensus on payment of annual fees by all partners, despite discounts being offered. Therefore integration would have meant that some partners would not join the organisation. Creating a legal entity for RIMA was also dismissed because it required not only time for preparation but also significant resources to operate thereafter. This would have required annual fees that were again not acceptable to partners.

RIMA has therefore chosen to move forward and achieve sustainability by creating a RIMA Alliance. This is inclusive and also viewed as convincing and complementary to existing organisations.

The Alliance is described in more detail in the following chapters, which describe the goals, way of working and how new partners can join the alliance.

SIMILAR INITIATIVES

PROJECT DEVELOPING CONCEPTS CLOSE TO DIHS

Certain RIMA partners are and have been involved in different European projects involving the creation of organisations and facilities dedicated to innovation, the deployment of robotics and the establishment of links with industry. These include e.g., competitions for FP7 EuRoC project (2014 -2018), "Robotics Innovation Facilities" or RIFs for FP7 Echord++[4] (2013-2019), "Competence Centres" (CCs) for H2020 HORSE (2015-2020) and ESMERA[5] (2018-2021) projects. Although CCs and RIFs can be seen as forerunners to the concept of DIHs, these organizations differ in the way their activity is structured. CCs and RIFs are not organized on the principle of services. The analysis of sustainability and business models conducted in these projects are general. The analysis presented in this book represents a further step in the study of sustainability.

I4MS

*I4MS, **ICT Innovation for Manufacturing SMEs**, is a European initiative **supporting manufacturing SMEs and mid-caps** in the widespread use of information and communication technologies (ICT) in their business operations. Under I4MS, SMEs can apply for technological and financial support to conduct experiments allowing them to test digital innovations in their business via open calls*[6].

I4MS organizes a community and working groups and facilitates informal connections between fellow innovators, entrepreneurs and professionals. The I4MS program funds projects supporting SMEs in manufacturing. Organizations involved form a community more than a network—they do not have the same structure as DIH networks and are not organized around an offer of services. Projects supported by I4MS like FP7 EuRoC project (2014 -2018) and H2020 HORSE (2015-2020) both for robotics, BEinCPPS[7] for cyberphysical systems and IoT, and PULSATE[8] for additive manufacturing have created capacities to support innovation and *Competence Centres*, however nothing compared to the RIMA network came out from these initiatives[9].

DIH NETWORKS

Five networks of robotics DIHs were run in parallel with RIMA: DIH^2[10] and TRINITY[11] for manufacturing, agROBOfood[12] for robots in agriculture and DIH-HERO[13] for robots in healthcare. Some RIMA partners are also partners in these projects, which enables us to share approaches. More generally, RIMA partners are sharing their views about sustainability through the RODIN[14] and DIHNET[15] CSAs. The model followed for the sustainability of these networks is not completely set. The TRINITY project implemented a sustainability model very close to that used in RIMA.

RIMA NETWORK SUSTAINABILITY – THE RIMA ALLIANCE

The sustainability of the network beyond the term of the RIMA project is organized in the form of an alliance: the *RIMA Alliance*. The RIMA Alliance focuses on issues of strategic importance for

4 http://echord.eu/
5 http://esmera-project.eu/welcome/
6 https://i4ms.eu/about/
7 http://www.beincpps.eu/
8 https://i4ms.eu/project/pulsate/
9 https://i4ms.eu/innovations/
10 http://www.dih-squared.eu/
11 https://trinityrobotics.eu
12 https://agrobofood.eu
13 http://dih-hero.eu
14 https://rodin-robotics.eu/
15 https://dihnet.eu/

the European economy and aims to promote knowledge creation and transfer between its members and third parties.

The vision of the RIMA Alliance is to be an actor that reinforces European leadership in I&M robotics by connecting European DIHs in an alliance that fosters cross border co-operation in robotics, artificial intelligence and big data in the field of inspection and maintenance for various industries. Research, technology & service providers, end users, investors, and certification bodies will be connected to the alliance through collaboration agreements, and will share challenges, solutions, information and training within the framework of the alliance.

The RIMA Alliance is not a legal entity. It relies on a contract between members that engages the members within a limited set of constraints. The main commitment for the members is to participate in one general assembly every six months on one day in a hybrid mode. To limit expenses, only resources and participation will be asked to the members—no fees will be required for membership.

Members

The founding members of the RIMA Alliance are the RIMA project partners. New members are welcome and will need to demonstrate an active interest in robotics for I&M and repair of infrastructure in any domain. They will need to have an interest in applied research, innovation and support for SMEs. Members can be

- **research and technology organizations or universities** interested in becoming a DIH, or existing DIH. These organizations will need to present their background in this robotics and in their activities related to I&M on infrastructure. They will need to present their offer in terms of services in the same way as current RIMA DIHs,
- **sectorial or professional associations** gathering industrial asset owners and asset operators in the different domains of application or technologies of interest for I&M of infastrucure,
- **clusters and facilitators** able to ease the connection with investors and business facilitators.

At the moment, SMEs working with I&M robotics will not be able to join the Alliance, however they are able to join the network through the community platform. In this way companies can receive any relevant information shared by the RIMA Alliance, for example regarding DIH services or events taking place.

Focus

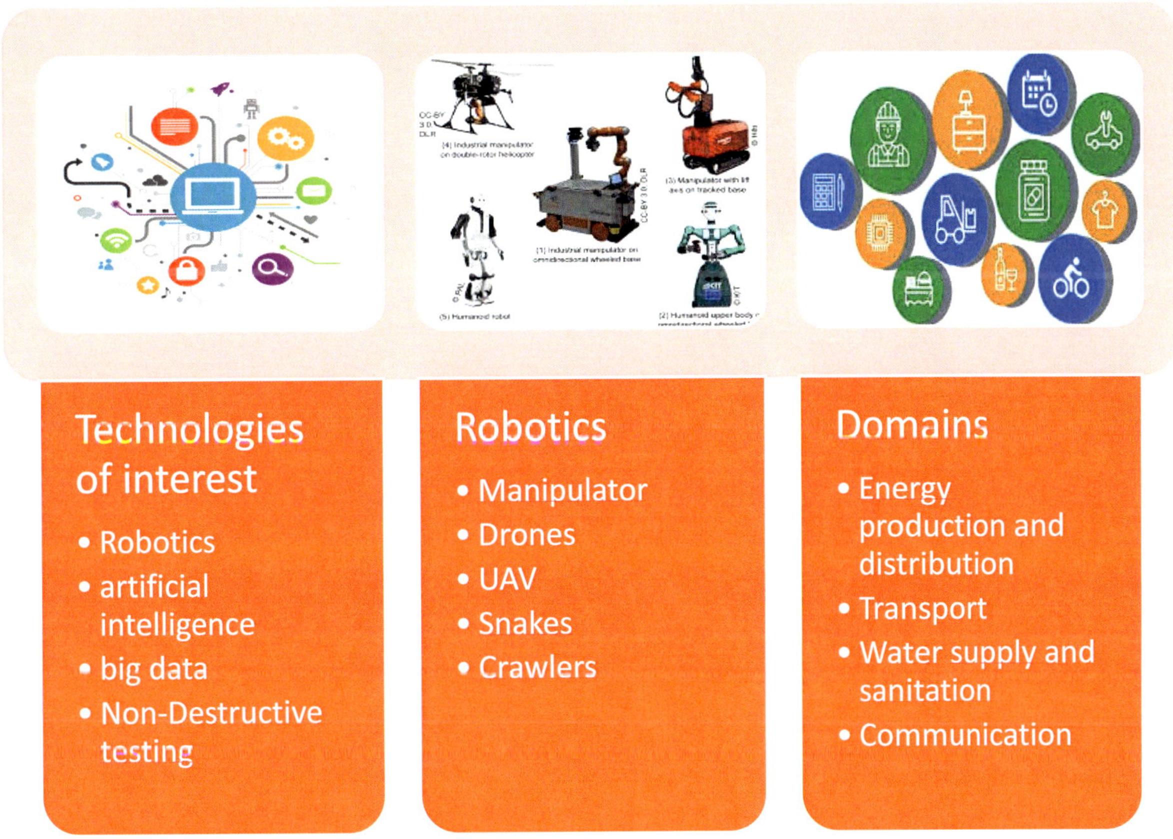

Figure 108: RIMA Network technologies, robotics and domains of interest

RIMA Alliance bodies and their organisation

The **RIMA Alliance General Assembly** (GA) is the ultimate decision-making body of the RIMA Alliance.

The **RIMA Alliance Board** is the entity managing day-to-day the RIMA Alliance. It comprises:

- a chair,
- a vice-chair,
- a secretary,
- a treasurer if necessary.

The board is elected during the General Assembly.

The RIMA Alliance General Assembly

Members

- One representative from each member.
- Each member has one vote, given through their representative.

Decisions

- The GA shall attempt to reach decisions by consensus
- A decision requires a quorum of 50 % of the members of the General Assembly.

Chair and vice chair

- Elections of the chair and vice-chair must be carried out freely and by secret ballot.

Meetings

- The GA shall meet at least twice a year, physically or virtually,
- The GA shall elect a chair, a vice-chair and a secretary out of the delegation of its members for a two-year term of office.

Activities and decisions of the GA

- activities/tasks/work to be performed;
- membership fee amount and due date;
- appointment of the chair, vice-chair, secretary and treasurer if needed;
- approval of candidates wanting to be become new members of the RIMA Alliance;
- identification of a breach by a Party of its obligations under this Agreement;
- termination of a Party's participation in the RIMA Alliance and measures relating thereto;
- appointment of other governance bodies or committees;
- appointment of the members of the External Stakeholders Committee if necessary;
- termination of the RIMA Alliance.

Agenda

- The Secretary shall prepare and send each member a written agenda
- Decisions will only be binding once the relevant part of the minutes has been accepted.

Finances

Parties must cover their own costs when implementing any action under this Agreement.

In order to implement the activities agreed by the GA, the Parties concerned can make specific financial arrangements between them. In such a case, they must consider the need to sign specific written agreements.

The General Assembly may decide to impose membership fees. This decision, the amount and due date, must be made unanimously.

Benefits of joining the RIMA Alliance

Each member has the opportunity to:

- Make use of the synergies between its members;
- Stimulate exchange between its members and third parties active in the relevant field;
- Organise dedicated workshop series,
- Encourage collaboration such as joint research actions between its members and third parties;
- Encourage publications, including joint publications, between its members;
- Propose education and training actions and foster secondment between its members;
- Propose state of the art studies;
- Foster creation of new skills and competences to answer stakeholders needs;
- Propose coordinated access to realistic inspection infrastructures;
- Work on organising and providing services through the Member in a coordinated manner;
- Support policy makers, EC and professional associations;

- Promote best practice in its domains of interest by developing new guidance documents,
- Increase the visibility and acceptance in the research community active in the field,
- Underpin regulation and standardisation through research, knowledge transfer and services,
- Work on gaining international leadership and recognition,
- Establish a comprehensive, longer-term infrastructure to respond to the needs,
- Support knowledge transfer, uptake, and industrial exploitation.

Engagement of organizations in the alliance

The process of joining the alliance is the following:

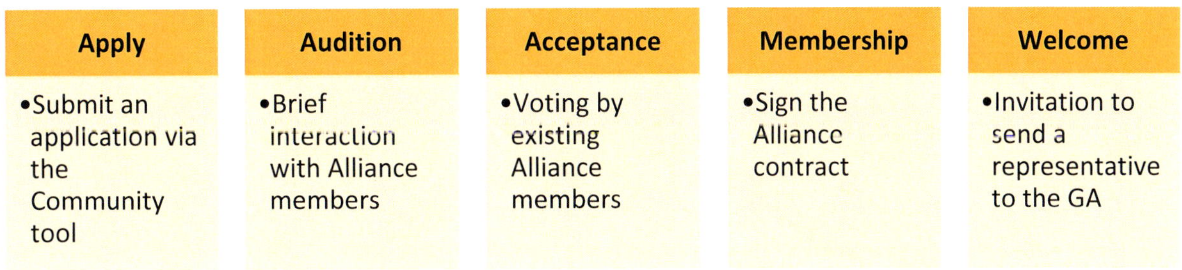

Figure 109: Process for joining the RIMA Alliance

Applying for Alliance registration

Each new DIH candidate will need to provide the following information on the community portal[16].

- Short description of the applicant.
- Relevance for I&M robotics.
- Domains covered.
- Robotics technologies investigated.
- Links, videos or similar to at least 3 projects or activities related to I&M robotics.
- For a DIH, describe at least 3 pieces of equipment, test facilities, or other infrastructure relevant for I&M robotics.
- Services offered relying on the model of the service catalogue.
- Ecosystem short overview, contacts with main actors.

The remainder of the joining process

- **Audition**: Brief interaction with Alliance members.
- **Acceptance**: Voting by existing Alliance members.
- **Membership**: Signature of the Alliance contract.
- **Dissemination**: Publication of the new member's information on the community portal and on social networks.
- **Welcoming**: The new member receives an invitation to send a representative to the General Assembly.

16 https://www.rimanetwork.eu

Relations with existing entities

The RIMA Alliance is complementary to existing organizations in robotics, AI, big data and infrastructure I&M. The Alliance will maintain connections with relevant existing organizations such as euRobotics[17], Sprint Robotics[18], adra[19], CLAIRE[20], BDVA[21], ELLIS[22], and EurAI23. Possible interactions include:

- Support for roadmapping, definition of strategic research agendas and definition of orientation.
- Support for organizing events.
- Connection between research and industry.
- Collaboration on marketing and outreach.

Figure 110: RIMA Alliance conections

Communication and dissemination tools

The community tool (Final URL currently being under discussions) was developed by Sprint Robotics during the RIMA project. It contains a vast amount of information about the RIMA network: descriptions of the DIHs, of the facilitators and sectorial associations member of the network. It contains information about the experiments monitored during the project as well as other information on innovation useful for inspection and maintenance stakeholders (see Figure 111)

17 https://eu-robotics.net/
18 https://sprintrobotics.org/
19 https://adr-association.eu/
20 https://claire-ai.org/
21 https://www.bdva.eu/
22 https://ellis.eu/
23 https://www.eurai.org/

Figure 111: RIMA Network community portal

To faciliate its evolution and managment, the community tool is being transfered from Sprint Robotics to euRobotics. This change of platform will not affect the tool's purpose and content. The new platform will include information on RIMA DIHs and their services, events, public documents and an option for SMEs to sign up to the network as associated members of the RIMA Alliance.

The Twitter and LinkedIn social media accounts will remain under the control of the RIMA Alliance.

Sustainability of the RIMA service offer

Our analysis of sustainability of the RIMA services offered by the DIHs was guided by the following postulates:

- In RIMA, the **sustainability of a DIH is the sustainability of the services** offered by this DIH. DIHs in RIMA are research organizations, universities, and institutions. Most of them are public organizations supported financially by states. All organisations involved as a DIH in RIMA have a long history and the sustainability of the organisation is often guaranteed by statute.
- **Services offered may differ from one DIH to the other**. Some DIHs may offer a service that others don't.
- The **business model ensuring the sustainability of a service can vary according to the service**. Some services can be offered for free, some require a fee to be paid.

With this in mind, the analysis of the sustainability of RIMA's DIHs consisted of studying the sustainability of each service for each DIH in the network. A thorough analysis of the services offered by each DIH is provided in the appendix. This appendix lists the services offered by each DIH and, for each service, we provide the following elements:

- **What** is the service providing? A brief description of each service offered by the DIH.
- **How** is this service provided? Is it a B2B event, a workshop, does it consist of information distributed on a web portal, for instance.
- **Who** in the DIH is providing the service?
- **Involvement** of the DIH: what effort is the DIH providing and what is the profile of people offering the service?
- Financial charge to the customer.

- **Funding**: how is the service supported financially by the DIH?
- **Contacts:** how to get access to the service.

CHAPTER 12

RESULTS AND POLICY RECOMMENDATIONS

CHRISTOPHE LEROUX, BRUNO SICILIANO AND ANIBAL OLLERO

Support for digitizing the European economy - value proposition for the DIHs in RIMA, ground truth

The idea of DIH networks comes from the Digitizing European Industry initiative launched by the European Commission in 2016[1]. Networks of DIHs and digital platforms were instruments foreseen by the EC to favour innovation, establish links between technology development organizations and industrial companies which would benefit from taking up these technologies.

These ideas came from the observation that Europe was very successful in supporting research that produced very good results but that this research output was slow to be deployed so it could benefit European Industry. Much progress still has to be made regarding innovation and support for innovation. DIHs were initially and are still expected to favour transfer of technologies developed within European research projects to industry. DIH networks in Europe were and are supposed to share their best practices to create real impact in all regions in Europe.

RIMA DIHs existed before the RIMA project. RIMA's DIH backbone are research organizations with long experience in robotics for inspection and maintenance of infrastructure. All of them have existing good connections with industry, with SMEs, in innovation, in supporting the creation of new products, often via spin-off or start-up companies. RIMA partners gathered to form a consortium and share their practices, the information they held to support innovation in their regions to improve their efficiency. RIMA finished after running 4 ½ years from 2019-2023.

The RIMA network is active. We selected 50 experiments (i.e. mini projects) with consortiums of SMEs that we supported financially, technically and to develop their business. We now have clearer information about the needs of those SMEs and their expectations regarding DIHs. In addition to the knowledge that came from our concrete experience in supporting SMEs since the beginning of RIMA, we also conducted a global market study. This study confirmed our expectations regarding the support that DIHs can bring to companies innovating I&M robots from a customer point of view, from the SMEs that engaged with RIMA DIHs.

All SMEs mentored in RIMA acknowledge the expertise of all RIMA DIHs. All SMEs are therefore keen to get advice about developing their technologies and innovative solutions. All have technical capacities and appreciate and look for the support and advice that DIHs can provide. The SMEs are very demanding with respect to advice about funding opportunities. They are also looking for suggestions about how to develop their businesses based on market data and looking for information that provides them with a better understanding of the ecosystems relevant to their work.

End users, asset owners and operators are unanimous in showing their interest in the interventions of the RIMA DIH network when monitoring and supporting SMEs in their innovation experiments. Robotics is complex, it is at the convergence of several technologies: mechatronics, control, perception, decision-making, AI, processes, and more—end users do not have the expertise to ensure that an idea can be developed into an applicable and effective solution to a particular inspection and

1 Digitizing European Industry https://digital-strategy.ec.europa.eu/en/library/roundtable-digitising-european-industry-commissioner-oettinger-september-2016

maintenance issue. The RIMA DIH network between them have vast technological expertise, enough to offer them a guarantee and get their trust in the chances of success of the experiments.

Recommendation 1: Maintain and support the concept of a DIH network as an instrument to favour innovation in I&M robotics.

Access to testing infrastructures

Access to realistic testing facilities was a recurrent demand from all experiments selected in the two RIMA open calls, for all domains, for all technologies, and all regions. All SMEs developing an application using a drone, for instance, asked for access to facilities where they could test their technology and measure the efficiency of their solution. All the experiments developing solutions for the nuclear domain asked for access to testing infrastructures allowing them to assess the manoeuvrability of their robots, the capacity to operate the robot from a distance or to demonstrate that their technology is applicable in a realistic environment. We could mention similar examples of expectations from all domains involving infrastructure I&M. A wider description will be provided in the deliverables related to the RIMA mentoring activities (RIMA WP6).

Access to testing facilities enables SMEs to save the time that would be needed to set up the experimental environments themselves, and makes more comprehensive testing possible. Time spent represents additional costs and time in the development of a solution. Realistic testing facilities independent from end users can, in the eyes of SMEs, be places where they can take measurement independently from the specific constraints of an end user's case. One expectation is that testing infrastructures can be places that help to certify that a solution conforms to regulations or a set of constraints, including safety considerations. Many end users require that the robot solutions provided are tested (and to some degree "in-house certified" on the end user's premises). Testing facilities could be in that respect a place helping to define standards, protocols, and provide a pathway towards certification.

Recommendation 2: Support the federation, development and networking of infrastructures for proving concepts in robotics for I&M in realistic conditions and certifying that the concepts respect regulations, safety, security, operational and environmental constraints.

EDIH a perspective for RIMA DIH?

In 2020, the European Commission launched the European Digital Innovation Hub (EDIH) concept. This concept is supported under the Digital Europe Programme. See also *Digital Innovation Hubs as policy instruments to boost digitalisation of SMEs*2 for further explanations. The service category breakdown in the EDIH differs slightly from the categorization used in DIHs. For consistency with the work on services already carried out and presented in RIMA, we have decided to keep the decomposition into categories that we have used since the beginning of the project. However, in order to illustrate the continuity between the concept of services in the RIMA network of DIHs and the one used in EDIHs, we will present in our conclusions the way in which this categorisation of services can be transposed from the notion of DIHs to that of EDIHs.

Some of the main differences between the concepts of EDIHs and of DIHs in RIMA are listed below (see page 20 of the policy instrument document mentioned above).

- In RIMA, DIHs are focused on robotics technologies, EDIHs cover several technologies, though each individual EDIH specializes.
- In RIMA, DIHs focus on inspection and maintenance of infrastructure, EDIHs cover all sectors, though again each individual EDIH may specialize.

2 https://publications.jrc.ec.europa.eu/repository/bitstream/JRC121604/final-onlineversion-dih-%28handbook%29.pdf

- In RIMA, DIHs focus on providing support to industry and in particular to SMEs to innovate and create new products and services. EDIHs are meant to support the uptake of existing techninologies.
- EDIHs were selected with the participation of Member States, whereas DIHs are part of the RIMA Alliance because of their existing expertise in inspection and maintenance and robotics.
- Some regions where DIHs are implemented will not set up a EDIH at all.

The future of RIMA DIHs does not lie in the EDIH concept.

Inspection and maintenance presents a specifc set of problems for robotics. Whatever the sector: energy production and distribution, water supply and sanitation, transport, the concerns about safety of operators, security of installations, the constraints about regulation, costs and time of operation are the same. Most of the time, human operators are needed to control that everything is going well. Tasks are not purely repetitive – adaptation is needed during the operation. I&M robotics is in that sense different from manufacturing robotics where robots can often just repeat operations automatically, most of the time in cages. The RIMA network involves organizations that have a long-term experience, background and strong expertise on the constraints of robot use in I&M. These organizations share the same «*language*», the same understanding of the needs of operators or owners of infrastructures.

Recommendation 3: Maintain the specifics of the concept of robotics DIHs focused on inspection, maintenance and repair of infrastructure, to take advantage of the strengths and complementarities of organizations sharing the same language, the same concerns, the same types of expertise and the same understanding of asset owners or asset operators to favour research and innovation.

DIH sustainability—some essential considerations

Building DIHs based around research organizations is an essential condition for the sustainability of a DIH. Experience shows, however, that establishing excellent liaisons with sector associations, financial advisors and clusters is also a key factor in the sustainability of a DIH. This is the approach followed in RIMA. Sectoral associations and clusters are key elements in linking up with industry in different sectors, avoiding the biases introduced by bilateral links, charting the way for multi-purpose solutions applicable to the wider European market and building perspectives beyond European borders. Sectorial associations enable us to address all their members through one contact point. They enable the identification of common challenges and common means of assessing the technologies, and can possibly arrange access to suitable testing infrastructures.

Connections with organizations expert in financing robot technologies and maintaining links with certification and regulatory bodies are also key features of successful DIHs. They facilitate understanding the constraints of the implementation of valid technological solutions in a wide range of applications. The members of the RIMA network are fully aware of these dimensions to success and most of them aim to develop them further in the future.

Recommendation 4: encourage and support the involvement of organizations able to involve all stakeholders in the network of DIHs. RIMA Alliance benefits from the participation of RTOs, universities providing technologies, sectorial associations able to identify user needs, and financial advisors.

Funding innovation

Funding to support innovation are obtained traditionally by DIHs from regional and national public funds, private funding from investors, or through bilateral contracting. Financial support, as was presented above, for innovation from the European Commission has had a substantial beneficial effect on the European economy. The funding mechanism based on FSTP proved to be very attractive to SMEs willing to innovate because of the simplicity and adequacy of the mechanism for testing proofs of concept. The mechanism enabled SMEs, asset owners, asset operators and DIHs to cover the risk when demonstrating that a concept was viable and effective to solve a problem in I&M. Most of the time, the challenges of I&M have been difficult to find relevant funding for developing and demonstrating a solution. Companies and private investors are hesitant to make the investment due to their lack of knowledge of the risks involved compared with the technological potential, and to their lack of expertise in estimating the chances for an experiment to reach the target goals.

RIMA DIHs were active before the project and will keep on being active beyond the end of the project. Their sustainability is ensured as each individual DIH is known and trusted by its customers, recognises the market potential, and many have statutory status. The experience gained via RIMA activities shows that the impact of the network of DIHs, the effectiveness of the support to industrials, and shortening of product development time, could be increased if it could develop certain services. Among the most frequently mentioned by SMEs are:

- The access to **independent, realistic test facilities.**
- The support for **test standardization.**
- The access to experts that can **provide advice on a range of relevant technologies.**
- The access to **funding**.
- The access to **market analysis**.

In order to keep providing the level of innovation activities/services reached during the RIMA project, we would need similar mechanisms for funding to the one already used.

These would be directions for a follow up to the RIMA network. This has a cost that will not be covered by the individual DIHs or the network without complementary funding.

Recommendation 5: organize, maintain, and possibly increase funding levels. Possibly use a mechanism similar to cascade funding to support innovation and cover the risks for proof of concepts for SMEs, asset owners, asset operators, and DIHs.

CHAPTER 13

CONCLUSIONS

CHRISTOPHE LEROUX, BRUNO SICILIANO AND ANIBAL OLLERO

At this tage, the RIMA Network has some major outcomes of the RIMA project that are worth to highlight:

1. the creation of the RIMA DIH network and now the RIMA Alliance,
2. demonstration of the effectiveness of the network of DIHs approach when supporting innovation in robotics for inspection and maintenance of infrastructure throught a new DIH catalogue of services on offer for European regions
3. about 50 upcoming technologies closer to market due to cascade funding and RIMA services
4. the RIMA community tool
5. various other reference documents and tools for the inspection and maintenance community.

RIMA DIH NETWORK

The creation of the RIMA Digital Innovation Hub (DIH) network gathers together organizations interested in innovation in robotics for inspection and maintenance (I&M) of infrastructure in different sectors. The organizations involved in the network include European sectorial associations and research organizations. The RIMA project demonstrated the effectiveness of the constitution of the network and the complementarity of the partners in covering the entire value chain. This complementarity enable us to match the needs of end users appropriately with the service offer from technology providers and test facility providers to accelerate innovation. The sectorial and professional associations facilitate understanding of sector challenges and market opportunities. They also enabled us to reach a vast number of industries sharing the same concerns. RIMA RTOs and universities formed DIHs providing innovative technologies which were the source of many new products and services.

Sectors covered in RIMA included energy production (nuclear, renewable, oil and gas) and distribution, water supply and sanitation, different modes of transport, and communication. RIMA experience shows that sharing challenges between these different sectors is highly beneficial in bringing commonalities to light, enabling the development of solutions applicable in several sectors and providing access to larger markets.

RTOs and universities were shown to be perfectly adapted to leading and being Digital Innovation Hubs, one-stop shops or doors for organizations having challenges to solve or wishling to get access to technologies not yet available on the market.

Although still needing consolidation and more practice, RIMA has also shown that presenting the offer of DIHs in terms of services can facilitate the dialogue between research and industry.

RIMA experience over the four+ years that it ran, demonstrated the interest in a network of DIHs from both DIHs and industry. Being part of a network enables the sharing of best practices. The network takes advantage of the complementarities between the partner DIHs in terms of technologies, in terms of sectors covered, and in terms of geographiy and language. Technologies used in I&M are very broad including aerial robotics, underwater, autonomous or semi-autonomous vehicles, arms, snake robots, teleoperated robots, cobots, human machine interfaces, AI, computer vision, non-destructive testing and much more. Sectors covered are also very broad. Despite all this variety, the uses of robots for I&M have a lot in common across the applications: tasks are not very repetitive, there is a necessity to be able to adapt to situations; safety and security constraints are very similar from one sector to the other; as are the costs and duration of interventions. These commonalities lie at the root of the unanimous interest for partners of RIMA to continue the DIH network.

The RIMA Alliance results from this unanimous interest and from the willingness of the RIMA project partners to extend their collaboration and keep the RIMA network alive beyond the term of the four-and-a-half-year project. Members of the RIMA Alliance sign a contract with light constraints. It does not need fees. It is not a legal structure. The main duty is to participate in a general assembly twice a year. Maintaining the DIH services to support SMEs and other companies in research and innovation, understanding the challenges of infrastructure I&M, sharing information about state of the art and about opportunities to collaborate are amongst the main motivations for the partners.

EFFECTIVENESS OF THE NETWORK

The number of products created by the SMEs supported by RIMA services illustrate the effectiveness of the network. Although the total impact of the experimentation is still under evaluation at this time (and complete understanding of the impact will take several more years), some indicators already show the success of network services in supporting innovation such as *Cost reduction, funds raised, products commercialised, services offered.*

In addition to these figures, feedback from SMEs that used RIMA services expressed their satisfaction with the support from the DIHs and the mentoring of project experiments. The excess of applications for cascade funding awards and the enquiries that are still coming in show that demand existed and still exists. Other organisations affected by RIMA work e.g. companies owning assets or interested in using the products being developed during the experiments also gave very positivie feedback about the RIMA concept and implementation.

The effectiveness of the network is also illustrated by the number of organizations willing to join the RIMA Alliance. At least seven non-partner research organizations have spontaneously expressed their interest in joining.

REFERENCE DOCUMENTS AND TOOLS FOR THE I&M COMMUNITY

One of the important outcomes of RIMA is the creation of the ***RIMA Community Tool***[1]. The community tool is accessible to all via the internet.2 It contains large amounts of information about the offer of the DIHs in the network, about the state of the art and technologies of interest for people and organizations interested in infrastructure I&M. Interested organizations can upload their own information. The community tool is now transfered to the euRobotics association for continuity and to reinforce the link and the will from the community to belong.

RIMA produced a **market study** to help understand the opportunities and challenges in the domain of I&M for infrastructure. This market study is now available to the community at large after registration through the community tool portal.

RIMA network partners elaborated a set of different **courses and trainings**. Some courses familiarize people with robotics in a hands-on approach, others are lectures to help people

1 Originally https://community.rimanetwork.eu/

2 At the momento of the edition of the book the final URL was under discusión. It should be available now at regular seach engines.

understand the technological offer inherent in I&M robotics, and some provided information about the funding available to support innovation.

RIMA partners published two documents to familiarize people with **cybersecurity issues in robotics** and one body of knowledge on regulations in robotics with a focus on I&M of infrastructure. These documents are also public and available for all thorough the community tool. Besides this book, several online publications and resources describe the results of the project.

THE FUTURE OF THE RIMA NETWORK

During the runtime of the project, RIMA demonstrated the effectiveness of and interest in our network of Digital Innovation Hubs as an instrument to support innovation in I&M of infrastructure. The network was set up well. Customers, SMEs, and large groups expressed their satisfaction many times on the results obtained with the support from RIMA. The services offered and the new products developed created positive impact on employment and on economic activity in Europe, demonstrating RIMA's effectiveness. We also succeeded in structuring an ecosystem that people are pleased to work with.

Now that the project is closed, the future lies in the RIMA Alliance. The Alliance is a light structure that has started to engage new members and which aims to maintain the connections between organisations that developed during RIMA. The RIMA Alliance will maintain services to SMEs to stimulate uptake of technologies, technology transfer, innovation and research. Focusing on services, on innovation, on robotics and on I&M of infrastructures, the RIMA Alliance is complementary to existing associations. It will work hand in hand with especially euRobotics, Sprint Robotics and ADRA.

CONTRIBUTORS

Christophe Leroux

Christophe Leroux received a PhD in Decision Theory from Sorbonne University in 1990. He is a senior researcher in AI for robotics and currently Manager of European Affairs at CEA. He represents CEA in several research projects. He has been involved in the creation of several networks about digitizing European industry and coordinates the project RIMA aiming at creating a network of Digital Innovation Hubs on the Inspection and Maintenance of Infrastructure.

Christophe Leroux is a member of the Board of Directors of the ADRA association and of the Board of Directors of the euRobotics association, both of which are involved in the AI, Data, and Robotics European research programme. He is scientific advisor for the APPROCHE association dedicated to robotics for handicapped people.

Christophe Leroux is the author of several scientific papers in conferences and journals about the applications of AI to robotics, on ethical, legal and socioeconomic issues in robotics, and on support for innovation in robotics.

Aneta Gałązka

Aneta Gałązka is an experienced Project Manager at FundingBox Accelerator. For the past 4,5 years she has been involved in highly innovative projects with SMEs, startups, and digital innovation hubs. Aneta has M.A. in Linguistics and a professional certificate in Coaching from Warsaw School of Economics.

Aneta Gałązka has managed more than 25 open calls, she was responsible in particular for: Open Call management, organising brokerage events, coordination of matchmaking, coordinating the online Jury Days, the selection process (coordination of evaluators' performance), community building, follow-up financial support to beneficiaries (accepting milestones, approvals of the payment), monitoring, measuring impact, and managing sustainability, moving to successful final reporting of the project. Her project experience - other than RIMA - includes: Ledger, DIH2, Tetramax, IMPACT Poland, AMULET, euROBIN, and TARGET-X.

Mariann Merz

Dr. Mariann Merz received her MSc in Aerospace Engineering from the University of Minnesota, Minneapolis, Minnesota, USA in 2001, and her PhD related to Unmanned Aerial Vehicle technology from the Department of Engineering Cybernetics at the Norwegian University of Science and Technology in 2022. Since February 2008 she has been a research scientist with SINTEF in Trondheim, Norway. Earlier work history includes working as an Autoflight Systems Analyst with the Boeing Company from 2001 to 2007. Her research interests include topics related to the design, path planning, control, and safety of autonomous vehicles and robots. She has authored/co-authored more than 10 scientific papers within the area of safety, robotics and autonomy.

Aksel Transeth

Dr. Aksel Transeth is a senior research scientist at the Department of Mathematics and Cybernetics at the SINTEF Digital institute in SINTEF in Norway and has a PhD in robotics and control from the Department of Engineering Cybernetics at NTNU, Norway. Dr. Transeth works with research projects and business development related to robotics and autonomy, with a particular focus on inspection and maintenance (I&M) robotics – from underwater applications to space applications.

Dr. Transeth co-initiated and leads a Norwegian industrial network on I&M robotics. He has authored/co-authored more than 36 scientific papers within the area of robotics and autonomy and has organized several national and internal workshops on robotics.

Claudio Paliotta

Claudio Paliotta received his M.Sc. in Mechanical Engineering from the University of Cassino and Southern Latium, Cassino, Italy, in 2013. In 2017 he received his Ph.D. in Engineering Cybernetics from the Norwegian University of Science and Technology, Trondheim, Norway. In 2022, he earned a M.B.A. from the Graduate School of Management of Politecnico di Milano, Italy. During his Ph.D., he carried out research on topics like path following control, trajectory tracking control, synchronization and coordination control of autonomous marine vehicles. He is currently working as Digitalization Specialist at Aker Solutions contributing to the digital transformation journey of the energy & utilities sector. Before joining Aker Solutions, he worked at SINTEF AS, one of the largest research centers in Europe, where he held different roles from research scientist to research manager. He has authored and co-authored 22 scientific publications that appear in international peer-reviewed conference proceedings and journals.

Tsampikos Kounalakis

Tsampikos Kounalakis is a robotics researcher at the Danish Technological Institute (DTI). He holds a B.Sc. (2011) in Applied Informatics and Multimedia from the Technological Educational Institute of Crete. He obtained a Ph.D. (2015) in 3D Computer Vision from Brunel University, United Kingdom. Tsampikos has been employed as a post-doctoral researcher at the Robotics, Vision and Machine Intelligence Laboratory (RVMI) and AAU Medialogy section in Copenhagen. He has been involved in various research projects and has organised several robot workshops. His research is in the area of machine learning, deep learning, mobile robotics, computer vision, and image processing.

Bruno Siciliano

Bruno is an Italian engineer, academic and scientific popularizer. He is professor of Control and Robotics at the University of Naples Federico II, Chair of the Scientific Council of the ICAROS Center, and Coordinator of the PRISMA Lab at the Department of Electrical Engineering and Information Technology. He is also Honorary Professor at the University of Óbuda where he holds the Rudolf Kálmán chair. He has received numerous international prizes, including the recent 2022 Engelberger Award for Education. Students in most prestigious universities study Robotics from his textbook. His Springer Handbook of Robotics received the Prose Award, the highest recognition for scientific books. "Keep the gradient" is the motto that Siciliano invented, meaning a constant search for new ideas and new solutions. A hymn to complexity to capture challenges and opportunities always in the name of the art of work & play. He is a passionate fan of Napoli football team, rock music, gourmet food and fine wines.

Jonathan Van der Meer

Jonathan van der Meer is Project Manager at PRISMA Lab in the Department of Electrical Engineering and Information Technology at University of Naples Federico II. He received the M.Sc. degree in Economics from the Free University of Amsterdam (1983). He is responsible for project planning, financial, administrative and legal issues as well as FSTP practices related to European projects.

Juan Antonio Pavón Losada

Currently embarked on a PhD In Digital communication, Pluralism and EU policy, Juan works for CREATE, Universita di Napoli, in RIMA and euROBIN as communication and dissemination lead. Juan has also worked as an EU project officer at elDiario.es/Andalucia, and as an independent EU policy analyst since 2013. He has participated in 10 other EU projects including I4MS and 5GCroco with MWC and Funding Box. He worked at the European Commission DG EMPL as International Policy Analyst. Further, he elaborated several research papers and opinion articles for Fundación Alternativas, El Pais, Diario de Sevilla. He directed the work on 4 books for RIMA, I4MS and Informe del Estado de Sevilla 2020.

Holding a Social and Labour Policy degree from Seville University (2008) his academic background includes a Communication Research master from US (2023), Political and Institutional Communication Master from Pompeu Fabra, (2018) and EU Public Affairs from Maastricht University (2013).

Maarit Sandelin

Maarit Sandelin currently works as the European Network Manager for robotic innovation at SPRINT Robotics. She holds a M.Sc. in Information Technology from the University of Oulu and earned her MBA from Durham University. She is an experienced project and product manager with various roles in consumer electronics and telecom ranging from technical to managerial to business development. She has organised several workshops, technology events and webinars on robotics.

Andreas Prodromou

Dr. ir. Andreas Prodromou is the Entrepreneur-in-Residence at the RCDS DIH. He is an engineer with an entrepreneurial spirit interested in activities at the intersection of technology and business. Andreas has more than two decades of experience in technological entrepreneurship in a wide range of capacities. His expertise includes strategy and business development, RTD, product development, technology transfer and management across various sectors (e.g., air, sea and ground transport, software, materials, and venture capital). Andreas has worked extensively with commercial entities of all sizes: from startups to multinationals, and has extensive experience working with non-profit entities (academia, national and international funding bodies, and technology transfer organisations). His activities are multinational, spanning three continents (Europe, the Americas and Asia). Andreas holds BS and MS degrees in Aerospace and Mechanical Engineering from Boston University (USA) and a Doctorate in Materials Engineering (Doctor in de Toegepaste Wetenschappen) from the K.U. Leuven, Belgium.

Savvas Loizou

Savvas Loizou received his Eng. Diploma and his PhD in Mechanical Engineering from the National Technical University of Athens, in 2000 and 2005, respectively. He has been a postdoctoral researcher at the GRASP Laboratory (University of Pennsylvania), a visiting lecturer at the Naval Architecture and Marine Engineering Dept. of NTUA, and an Assistant Professor at the Mechanical Engineering Department of Frederick University, Cyprus. He joined the Department of Mechanical and Materials Science and Engineering of Cyprus University of Technology in January 2011, establishing the Robotics, Control, and Decision Systems Laboratory (RCDS). He is currently an Associate Professor and since 11/2020 the Department's chairman. His research interests span theoretical and applied issues involving robotics and autonomous dynamic systems. His application domains include underwater, underground, mobile and aerial robotics. He is a member of IEEE and the Technical Chambers of Cyprus and Greece.

Sotiris Makris

Dr. Sotiris Makris obtained his Diploma (Degree Master of Science) in Mechanical Engineering and Aeronautics from the University of Patras, Greece and a PhD in Engineering from the same place. He worked as a research associate for the Laboratory for Manufacturing Systems and Automation (LMS) in the Department of Mechanical Engineering and Aeronautics from September 1997 to the present. His main research interests are focused on the field of robots, automation and virtual reality in manufacturing. He speaks fluent English. He has been involved in more than fifty (50) RTD projects funded by the EC, acting as senior researcher, project and technical manager contributing to the research and technological advancement of the Laboratory for Manufacturing Systems and Automation. Major EC funded projects that he managed are the following: SHERLOCK, ESMERA, SHAREWORK, MyCar, THOMAS, SYMBIO-TIC, VERSATILE, AUTORECON, X-act, INTERACT, Sense&React, ROBO-PARTNER, MARVIN, Intermar, Syrios, e-Custom. He has more than 150 scientific articles published and he has been an associate editor for two (2) international scientific journals and a reviewer for thirty (30) international journals, in a large number of international conferences and guest editor in two (2) international scientific journals. He organized the Second CIRP Web Conference.

George Michalos

Dr. George Michalos obtained his Diploma in Mechanical and Aeronautics Engineering and a PhD in Engineering from the Department of Mechanical Engineering and Aeronautics, University of Patras. He worked as a research engineer for the Laboratory of Manufacturing Systems and Automation (LMS) from 2006 to 2012 and is currently employed as a research project manager in LMS (since 2013). He speaks fluent English and French. His main research interests are focused on advanced tools for the design and control of highly flexible production systems and on engineering research for industrial automation and robotics applications. He has performed research in the areas of decision-making for the automated design of flexible manufacturing systems, of planning and control of flexible manufacturing systems, of reconfiguration of assembly systems, of application of RFID (Radio Frequency Identification) technology for product tracking and automated operator support in assembly lines, and more.

Panagiotis Karagiannis

Panagiotis Karagiannis is an Electrical and Computer Science Engineer from the University of Patras. He has worked over 10 years for LMS, focusing his research activities on HRI, perception technologies and flexible assembly cells. He has been involved in a number of EC funded projects both for manufacturing research (AUTORECON, X-ACT, ROBO-PARTNER, THOMAS, VERSATILE) as well as in FSTP-based projects such as RIMA (technical manager), TRINITY (education manager) and MASTER (project coordinator). Topics that he has worked on as a research engineer are HRI, using Augmented Reality technologies, robot controllers based on the ROS platform, safety concepts based on safety cameras and other safety sensors, computer vision algorithms both for 2D-vision and 3D-vision systems used for object and human recognition, flexible mechatronic implementations, hardware configuration, and software development tailored to industrial needs.

Sotiris Aivaliotis

Sotiris Aivaliotis has worked as a research engineer at LMS from 2016 to the present day. His main research interests and work focus on the fields of robotics, design and simulation with CAD tools (e.g. CATIA, DELMIA, Visual Components, Process Simulate, Solidworks), design of Human-Robot Collaborative (HRC) assembly lines, Robot Operating System based platform developments, industrial safety systems for HRC applications and Augmented Reality applications. Over the past years he has been involved in several projects funded by the EU, including THOMAS, ODIN, ROBO-PARTNER, RIMA, TRINITY, VERIFY.

Krzysztof Lipiec

Krzysztof Lipiec is director of commercialization department at the Łukasiewicz-PIAP Institute. He is responsible for the implementation of research results, mostly in the area of robotics, into business practice and supervises the process of obtaining funding from public sources. He is the initiator and leader of European programs supporting digitization supported from H2020 and Digital Europe programs (DIH, EDIH). He represents the community of the European Institute of Innovation and

Technology EIT-Manufacturing in Poland. He participated in the development and was responsible for the implementation in Poland of the strategic Advanced Manufacturing evaluation and planning methodology ADMA under the service contract European Commission Executive Agency for Small and Medium-sized Enterprise (EASME). He oversees the work of the Horizon Europe Industry contact point for Industry 4.0. and the European Digital Innovation Hub EDIH Mazovia.

Jakub Główka

Jakub Główka is head of the Security and Defense System division at Łukasiewicz-PIAP. He is responsible for the implementation of research projects mostly in the area of cyber-physical systems with special emphasis on new use cases for robotics. He successfully manages a research group very active in international cooperation within European programs supporting information technologies including H2020, Horizon Europe, Digital Europe, EDA, ESA and NATO programs. He has participated in more than 20 research projects. He is one of the initiators of European programmes supporting digitization supported via H2020 and Digital Europe programs (DIH Networks, EDIH). He works on a technical level with local SME representatives to introduce emerging cyber-physical technologies to local markets and clients.

Anibal Ollero

PhD of Engineering and Professor of Robotics at the University of Seville, he has been a professor at the universities of Malaga and Santiago de Compostela, with research stays at the French CNRS and the Robotics Institute of Carnegie Mellon University in the USA. All focusing on aerial robotics and unmanned aerial vehicles. In Seville, Anibal directs the Robotics, Vision and Control Laboratory (GRVC) research group and is also an adviser to the Center for Advanced Aerospace Technologies (CATEC). He is the author of more than 850 publications, including nine books, being the world's most published author on unmanned aerial vehicles and aerial robotics. He has directed over 47 doctoral theses. He has received more than 28 prizes and distinctions, including the Overall ICT Innovation Radar Award from the European Commission in 2017, the King Jaime I Award in New Technologies in 2019, and the Leonardo Torres Quevedo National Research Award in 2021.

Begoña Arrue Chiquinquira

Associate Professor Arrue (Dep. System Engineering, University of Seville) holds a BSc in Computer Science and an MSc and a PhD in Electrical Engineering. Arrue is a lecturer for undergraduate and postgraduate engineering students and is an active researcher with the GRVC Lab in Seville. By integrating multidisciplinary approaches, her research aims to devise novel bio-inspired robotic paradigms focused on aerial soft robotics. Arrue's areas of expertise also include AI in robotics, perception, manipulation and planning, and intelligent systems. She has published 70+ peer-reviewed articles and has participated in 35+ research projects (national and international). Dr. Arrue has led 3 EU-funded grants. At present, Arrue is involved in several research initiatives and is co-PI of the European projects RAICAM and RIMA. She has an extensive industrial engagement track record. She is also an IEEE Senior member and has been a scientific assessor for international journals and national and international research programs.

José Luis Andrade Pineda

José Luis Andrade Pineda is Telecommunications Engineer (2000) and holds a Master Degree in Organization Engineering (2009) from the University of Seville, where he obtained the title of Doctor of Engineering in 2015. Member of the Research Group TEP-134 "Industrial Organization", José has experience in innovation processes by incorporating technology into industrial environments and services (2007-2017) and has published 15 articles in journals indexed in the ISI-JCR. He has participated in change management procedures, drafting technical instructions for transformation of operations, ITIL and PMBok methodologies, and communications deployment field work. Since 2018 he has been working as a post-doctoral researcher hired by GRVC at the

University of Seville. His main focus is on the inclusion of aerial robotics solutions in I&M activities of large industry (petrochemical, electrical, nuclear), with special attention to functional safety and the redefinition of operating processes derived from its automated or semi-automated operations.

APPENDIX

TABLES OF SERVICES OFFERED BY THE RIMA DIHS

This appendix expands the services resumés given in section 6. It gives a snapshot of the service offering from each DIH as of March 2023. These services will alter in response to demand.

Robotics DIH Spain

The GRVC lab from the University of Seville leads the Robotics DIH Spain with the active participation of FADA CATEC and the support of Technological Corporation of Andalusia (CTA).

BREAKDOWN OF SERVICES

1. Ecosystem services

Awareness creation	
Category	Ecosystem service, Community building, Ecosystem learning.
Description of activity, value proposition (what?)	Robotics awareness presentations for decision-makers and suppliers, and promotion of working groups.
Format (how?)	Webinars/workshops/conferences where the DIH is invited to present, along with dissemination of the conclusions/recommendation of the working groups (e.g., new regulatory frameworks on UAS).
Profile of DIH contact (who?)	Senior technical experts.
Involvement, effort from DIH	Presentation by one technical expert.
Financial charge for customer	Free.
Funding, business plan	Subsidized by the DIH or supported by the inviting organization.
Dissemination	
Category	Ecosystem service, Community building, Promotion of robotics in I&M.
Description of activity, value proposition (what?)	Communication of the services offered by the DIH. Dissemination of results and key facts in local language and in English.
Format (how?)	Information distributed through traditional mass media (press) but also digital ones e.g. magazines, web sites, social media, newsletters, email.
Profile of DIH contact (who?)	Interviews of technical experts, experts in communication.
Involvement, effort from DIH	Involvement of the communication teams belonging to the organizations involved in the DIH.
Financial charge for customer	Free.
Funding, business plan	Subsidized by the DIH.

Ecosystem building	
Category	Ecosystem service, Community building.
Description of activity, value proposition (what?)	Active participation in forums spreading the technology and other relevant aspects for I&M, allowing contact with professional associations, sectorial associations, certification bodies, public and private investors in region and in Europe.
Format (how?)	Peer-reviewed scientific works which first are presented at ERF, ICRA, IROS, ICUAS,… and then become reference information provided through the DIH web portal and via DIH contact points.
Profile of DIH contact (who?)	Business facilitators, technical experts.
Involvement, effort from DIH	Actions to gain visibility on the ecosystem. By means of a kind of Technological Watch, the business facilitator in the DIH disseminate among networks, clusters, associations, and federations to which the organizations of the DIH belong and/or has contacts with.
Financial charge for customer	Free.
Funding, business plan	Subsidized by the DIH. Potential economic upside for DIH in generating projects which involve the DIH.
Brokerage, Matchmaking	
Category	Ecosystem service, Community building, Ecosystem learning.
Description of activity, value proposition (what?)	Organisation of meetings and workshops between players in the value chain to bring demand and offer sides together to share ideas and analyse opportunities for collaboration.
Format (how?)	Minimum one event per year at a relevant location (e.g., a petrochemical plant, a furniture, etc.), our own facilities (e.g. an outdoor area wherein a demonstration activity is presented to the visitors) or online with presentations from stakeholders, networking lunch (if not digital), dialogue B2B, tour of the DIH if visit on site.
Profile of DIH contact (who?)	Business facilitators, I&M technical experts, experts on the technology being demonstrated, etc.
Involvement, effort from DIH	One-day event with participation of 3 to 9 experts from the DIH, supply of communication material.
Financial charge for customer	Free.
Funding, business plan	Subsidized by the DIH.

Market research, market awareness, market/technology trends	
Category	Ecosystem service, Strategy development.
Description of activity, value proposition (what?)	Provision of contact to experts in market analysis, market data held in the network and at the DIH at European and international level.
Format (how?)	Conference call, meeting with business developers at the DIH.
Profile of DIH contact (who?)	Contact to experts in market analysis.
Involvement, effort from DIH	1-2 hour with experts.
Financial charge for customer	Yes.
Funding, business plan	Subsidized by the DIH.
Mentoring by technical and application experts	
Category	Ecosystem service, Strategy development.
Description of activity, value proposition (what?)	One-to-one follow-up with relevant entities (e.g. asset owners) to concretize the potential use of robotics and automation to create value in their organization.
Format (how?)	Conference calls, bilateral meetings, meetings, or trials on customer's premises.
Profile of DIH contact (who?)	Technical and application experts.
Involvement, effort from DIH	A mid-day of a set of 3-6 experts from the DIH.
Financial charge for customer	Free.
Funding, business plan	Subsidized by the DIH (as part of investigating project opportunities) or cost per hour for larger efforts (when the action has specifically requested by the relevant entity).
Resources and knowledge sharing	
Category	Ecosystem service, Ecosystem learning.
Description of activity, value proposition (what?)	Informing about and organizing access to resources, databases, seminars, workshops, conferences, field trips, newsletters, and other information of interest.
Format (how?)	Reference information provided through the DIH, newsletters and RIMA network web portal. Conferences, field trips, workshops organized by the DIH or in cooperation with partners.
Profile of DIH contact (who?)	Communication team, experts.
Involvement, effort from DIH	Involvement of the communication teams of CTA, CATEC and USE, the organizations involved in the DIH.
Financial charge for customer	Free (information sharing, some conferences) or paid (e.g., conference fee)
Funding, business plan	Subsidized by the DIH. Mixed public and private funding.

DIH network promotion, Promotion of the offer	
Category	Ecosystem service, Ecosystem learning
Description of activity, value proposition (what?)	Promote both the DIH and its members.
Format (how?)	Webpages, newsletters. Participation in roadshows, exhibitions, dissemination, and other activities according to customer interests (e.g., suppliers, end users).
Profile of DIH contact (who?)	Communication, business, and technical experts.
Involvement, effort from DIH	Involvement of the communication teams and technical experts.
Financial charge for customer	Free or paid.
Funding, business plan	Subsidized by the DIH. Mix of public and private funding.

2. **Business services**

Connect to funding sources	
Category	Access to financing, Incubator/accelerator support.
Description of activity, value proposition (what?)	Connect businesses and public organizations to relevant funding opportunities including entrepreneurship. Lead or contribute to matching for the funding opportunities. Business development as part of our service.
Format (how?)	Webinars, meetings, conference calls.
Profile of DIH contact (who?)	Business developer, technical expert.
Involvement, effort from DIH	Meetings with experts in funding for research, development, and innovation in robotics.
Financial charge for customer	Involvement in proposal/funding investigating phase is free for everyone, but further services (e.g. proposal writing) for CTA partners only.
Funding, business plan	Subsidized by the DIH when part of a limited-support offer (e.g., a 1–2-hour information meeting). Private funding when the DIH is involved only in the sales phase and not in the potential subsequent project.
Hands-on training	
Category	Skills and education.
Description of activity, value proposition (what?)	Training of technical and scientific personnel.
Format (how?)	Courses given by DIH personnel. On demand and regular.
Profile of DIH contact (who?)	Technical experts.
Involvement, effort from DIH	1-2 day events and shorter events, supply of training reference material.
Financial charge for customer	Yes.
Funding, business plan	Flat fee per seminar/training course.

Research, development and innovation opportunities	
Category	Project development, Access to finance.
Description of activity, value proposition (what?)	Advising on technological research, development, and innovation activity. Elaboration of an action plan for RDI with milestones, deliverables, agreement on the respective role of stakeholders and on the final delivery. Basis for initiating, e.g., bilateral projects between industry and RTOs in the DIH, or as basis for proposal writing to secure funding.
Format (how?)	Meetings, conference calls.
Profile of DIH contact (who?)	Business developer, technical expert.
Involvement, effort from DIH	Meetings with experts in research, development and innovation in robotics.
Financial charge for customer	Free if the activity can lead to project for the DIH, otherwise paid.
Funding, business plan	Subsidized by the DIH or private funding.
Hosting spin-off and start-up companies	
Category	Offering housing.
Description of activity, value proposition (what?)	Not directly provided by the DIH, but there is a service from CATEC and CTA that redirects to the Aerospace Business Incubator of the Seville Chamber of Commerce (AEROPOLIS).
Format (how?)	On demand.
Profile of DIH contact (who?)	Business facilitator.
Involvement, effort from DIH	Office space, workspace, located in DIH premises, internet infrastructure, lab facilities.
Financial charge for customer	Free.
Funding, business plan	Subsidized by CTA and CATEC.

3. Technology services

Research and development collaboration	
Category	Strategic RDI.
Description of activity, value proposition (what?)	Research aiming to reach TRL 5, removal of doubts, specifications, research and development, validation in laboratory.
Format (how?)	Collaborative research with several partners in Spain, mainly on avoiding risks in I&M actions, for 2 years or more.
Profile of DIH contact (who?)	Project manager, technical experts.
Involvement, effort from DIH	Support from experts in the submission of European or national proposals. Long term participation (over 2 years) of several experts from the DIH for technical research and development in robotics.
Financial charge for customer	Free or paid and/or via in-kind from the industry.

Funding, business plan	Funded by European or national research programmes, mix of public and private funding, or solely private funding (addressing challenges brought by a private asset owner).
Innovation projects: technology demonstrators, proof of concepts	
Category	Contract research.
Description of activity, value proposition (what?)	Research and development aiming to reach TRL 6, application of existing technologies, validation in laboratory.
Format (how?)	Bilateral or collaborative projects from one to a few years.
Profile of DIH contact (who?)	Project manager, technical experts.
Involvement, effort from DIH	Support from experts in the submission of European or national proposals. Participation of several experts from the DIH for technical research and development in robotics, mainly as consultants for integration processes and low-level features at UAS systems.
Financial charge for customer	Yes.
Funding, business plan	Funded by bilateral contracts, European or national innovation programmes, private investors.
Proof of concept, prototyping, commercialization/industrialization, small series production	
Category	Technical support on scale-up.
Description of activity, value proposition (what?)	Demonstration in real environment or in realistic conditions. Adaptation of existing technologies to specific use cases.
Format (how?)	Bilateral projects from one to several months.
Profile of DIH contact (who?)	Technical experts.
Involvement, effort from DIH	Supervision of final manufactured works according to Proof-of-Concepts formerly defined under contract. Support from experts in the DIH for attaining a commercial product on the in-place Proof-Of-Concept.
Financial charge for customer	Yes.
Funding, business plan	Funded by bilateral contracts.
Lab facilities, equipment for lease/use (sensors, robots, robot tooling, etc.), rapid prototyping	
Category	Provision of technological infrastructure.
Description of activity, value proposition (what?)	Demonstration with equipment held in the facilities (both, indoor and outdoor) or deployment of robotics solutions (on-board sensors, actuators, aircrafts, etc.) at customer site. Provision of test benches (for actuation, perception, etc.) whether in the VICON system in CATEC or in the OPTITrack system in USE. Provision of large-scale flying areas (e.g., test area in the experimental test flight center ATLAS).

Format (how?)	Conducting experiments with the assistance of DIH personnel.
Profile of DIH contact (who?)	Technical experts.
Involvement, effort from DIH	Personnel from the DIH needs to be involved in the settings for ensuring proper exploitation of the facility's capabilities. A proper testing plan needs to be designed and agreed in advance.
Financial charge for customer	Yes, financial charge based on daily rate and depending on the personnel involved, equipment used at the DIH and adaptation needed.
Funding, business plan	Funded by bilateral contracts, European or national innovation programmes, investors.
Certification, product demonstration & testing, comparison of products	
Category	Testing and validation.
Description of activity, value proposition (what?)	For testing and validation of product functionalities, CATEC and USE hold and can offer access to a set of dedicated infrastructures for assessment in realistic conditions (e.g., outdoor/indoor monitoring of multi-run flying experiments or long-range flights). The DIH can also organize access to infrastructure in other application sectors, through its access to other facilities at USE.
Format (how?)	Conducting experiments with the assistance of DIH personnel following evaluation protocols. Provision of metrics and tools for benchmarking the utility, usability, effectiveness.
Profile of DIH contact (who?)	Technical experts.
Involvement, effort from DIH	Support from subject matter experts, whether from the DIH or from our network contacts whose involvement helps to ensure the maturation of solutions.
Financial charge for customer	Yes.
Funding, business plan	Funded by bilateral contracts, European or national innovation programmes, investors.

C.R.E.A.T.E (Italy)

C.R.E.A.T.E. is both a legal entity and a consortium comprising 6 universities, ABB-Ricerca and Ansaldo-Energia.

BREAKDOWN OF SERVICES

1. Ecosystem services

Awareness creation	
Category	Ecosystem service, Community building.
Description of activity, value proposition (what?)	Robotics awareness course for decision-makers.
Format (how?)	Workshop or webinar organized by DIH or to which DIH is invited.
Profile of DIH contact (who?)	Technical experts.
Involvement, effort from DIH	Presentation by representatives.
Financial charge for customer	Free.
Funding, business plan	Subsidized by the DIH.
Dissemination	
Category	Ecosystem service, Community building.
Description of activity, value proposition (what?)	Communication of the key activities of the DIH. Dissemination of results and key facts from European and national projects where CREATE is involved.
Format (how?)	Information distributed through e.g., web sites, social media, newsletters, email, newspapers and TV presentations and interviews.
Profile of DIH contact (who?)	DIH representatives supported by experts in communication.
Involvement, effort from DIH	Involvement of the representatives belonging to the organizations involved in the DIH.
Financial charge for customer	Free.
Funding, business plan	Subsidized by the DIH.
Ecosystem building	
Category	Ecosystem service, Community building.
Description of activity, value proposition (what?)	Putting people in touch with professional, sectorial associations certification bodies, public and private investors in region and in Europe.
Format (how?)	Possible contact with experts.
Profile of DIH contact (who?)	Business facilitators, technical experts.
Involvement, effort from DIH	Access to networks, clusters, associations, and federations to which the organizations of the DIH belong.
Financial charge for customer	Free.
Funding, business plan	Subsidized by the DIH.

Brokerage, Matchmaking	
Category	Ecosystem service, Community building.
Description of activity, value proposition (what?)	Organization of meetings between organizations in the value chain to bring demand and offer sides together to share ideas and analyse opportunities and strategies.
Format (how?)	Two events per year at the DIH in collaboration with DIH Campania or online with presentations from stakeholders and dialogue B2B space.
Profile of DIH contact (who?)	Business facilitators, technical experts.
Involvement, effort from DIH	Half-day event with participation of 3 to 4 experts from the DIH and DIH Campania, supply of communication material.
Financial charge for customer	Free.
Funding, business plan	Subsidized by the DIH and DIH Campania.
Technical consultancy	
Category	Ecosystem service, Strategy development.
Description of activity, value proposition (what?)	Advise on potential uses of robotics and automation.
Format (how?)	Bilateral contacts with organizations.
Profile of DIH contact (who?)	Technical and application experts.
Involvement, effort from DIH	Individually based.
Financial charge for customer	In this stage free, paid if it leads to contracts.
Funding, business plan	Subsidized by the DIH.
Resources and knowledge sharing	
Category	Ecosystem service, Ecosystem learning.
Description of activity, value proposition (what?)	Informing about and organizing access to resources, databases, seminars, workshops, conferences, field trips, newsletters, and other information of interest.
Format (how?)	Reference information provided through the DIH and RIMA network web portal. Access to experts for advanced consultancy.
Profile of DIH contact (who?)	Project managers and senior researchers.
Involvement, effort from DIH	Involvement of the communication experts of the organizations involved in the DIH.
Financial charge for customer	Free or paid.
Funding, business plan	Subsidized by the DIH.

DIH network promotion, Promotion of the offer	
Category	Ecosystem service, Strategy development.
Description of activity, value proposition (what?)	Promote both the DIH and its members.
Format (how?)	Organization of roadshows, exhibitions, dissemination, and other activities according to customer interests (e.g., suppliers, end users).
Profile of DIH contact (who?)	Project managers and senior researchers.
Involvement, effort from DIH	Involvement of the communication teams belonging to the organizations involved in the DIH.
Financial charge for customer	Free.
Funding, business plan	Subsidized by the DIH.
Resources and knowledge sharing	
Category	Ecosystem service, Ecosystem learning.
Description of activity, value proposition (what?)	Informing about and organizing access to resources, databases, seminars, workshops, conferences, field trips, newsletters and other information of interest.
Format (how?)	Reference information provided through the DIH and RIMA network web portal. Access to experts for advanced consultancy.
Profile of DIH contact (who?)	Project managers and senior researchers.
Involvement, effort from DIH	Involvement of the communication experts of the organizations involved in the DIH.
Financial charge for customer	Free or paid.
Funding, business plan	Subsidized by the DIH.

2. Business services

Business development	
Category	Incubator/accelerator support.
Description of activity, value proposition (what?)	Dedicated to spin-off and start-up companies but implemented directly by the local organizations belonging to the DIH via a local autonomous organization.
Format (how?)	Individual contracts.
Profile of DIH contact (who?)	Experts in business creation.
Involvement, effort from DIH	Meetings with experts in business development.
Financial charge for customer	Depending on each member organization.
Funding, business plan	To be developed in collaboration with the interested companies.
Hands-on training	
Category	Skills and education.
Description of activity, value proposition (what?)	Training of operators, raising awareness about teleoperation, and robotics.
Format (how?)	Courses and hands-on training on equipment available at the DIH. On demand.
Profile of DIH contact (who?)	Technical experts.

Involvement, effort from DIH	One day with participation of experts, supply of training reference material.
Financial charge for customer	Services against payment.
Funding, business plan	Rates according to the trainer's profile.
Research, development and innovation opportunities	
Category	Project development.
Description of activity, value proposition (what?)	Dedicated to organizations interested in collaborative research.
Format (how?)	Meetings, conference calls.
Profile of DIH contact (who?)	Business developer, technical expert.
Involvement, effort from DIH	Meetings with experts in research, development, and innovation in robotics.
Financial charge for customer	Free.
Funding, business plan	Subsidized by the DIH.
Hosting spin-off and start-up companies	
Category	Offering housing.
Description of activity, value proposition (what?)	Service is provided by the individual CREATE member universities.
Format (how?)	On demand.
Profile of DIH contact (who?)	Business facilitator.
Involvement, effort from DIH	Office space, workspace, internet infrastructure.
Financial charge for customer	Yes.
Funding, business plan	Rates according to needs.

3. Technology services

Research and development collaboration	
Category	Strategic RDI.
Description of activity, value proposition (what?)	Research development, removal of doubts, specifications, research and development, validation in laboratory.
Format (how?)	Collaborative research with several partners in EU.
Profile of DIH contact (who?)	Project manager, technical experts.
Involvement, effort from DIH	Support from experts in the submission of European or national proposals. Long term participation (over 2 years) of several experts from the DIH for technical research and development in robotics.
Financial charge for customer	Free.
Funding, business plan	Funded by European or national research programmes.
Innovation projects: technology demonstrators, proof of concepts	
Category	Contract research.
Description of activity, value proposition (what?)	Research and development aiming at improving an application of existing technologies, validation in laboratory.
Format (how?)	Bilateral projects from one to a few years.
Profile of DIH contact (who?)	Project manager, technical experts.

Involvement, effort from DIH	Support from experts in the submission of European or national proposals. Long term participation (over 2 years) of several experts from the DIH for technical research and development in robotics.
Financial charge for customer	Yes.
Funding, business plan	Funded by bilateral contracts, European or national innovation programmes, private investors.

Lab facilities, equipment for lease/use (sensors, robots, robot tooling, etc.), rapid prototyping	
Category	Provision of technological infrastructure.
Description of activity, value proposition (what?)	Demonstration with equipment held in the lab. Realization of interfaces with DIH equipment. Provision of test benches (for actuation, perception, etc.)
Format (how?)	Conducting experiments with the assistance of CREATE staff.
Profile of DIH contact (who?)	Technical experts.
Involvement, effort from DIH	Participation of several experts from the DIH for maturation.
Financial charge for customer	Yes, financial charge based on individual agreement and depending on the personnel involved, equipment used at the DIH, and adaptation needed.
Funding, business plan	Funded by bilateral contracts, European or national innovation programmes, investors.

Cyprus University of Technology

BREAKDOWN OF SERVICES

1. Ecosystem services

Awareness creation	
Category	Ecosystem service, Community building.
Description of activity, value proposition (what?)	Open days, STEM education activities, euRobotics week, workshops.
Format (how?)	Workshops, open days.
Profile of DIH contact (who?)	DIH technical experts.
Involvement, effort from DIH	Presentations by DIH technical experts.
Financial charge for customer	Free.
Funding, business plan	Subsidized by the DIH, research projects.
Dissemination	
Category	Ecosystem service, Community building.
Description of activity, value proposition (what?)	Communication of the offer of services of the DIH. Dissemination of results and key facts in local language and in English.
Format (how?)	Information distributed through e.g., web sites, social media, newsletters, email, infodays, workshops.
Profile of DIH contact (who?)	Technical and business experts.
Involvement, effort from DIH	Organization, communication, execution.
Financial charge for customer	Free.
Funding, business plan	Subsidized by the DIH, research projects.

Ecosystem building	
Category	Ecosystem service, Community building.
Description of activity, value proposition (what?)	Putting people in touch with professional and sectorial associations, certification bodies, public and private investors in the region and in Europe.
Format (how?)	Reference information provided via DIH contact points.
Profile of DIH contact (who?)	Business facilitators, technical experts.
Involvement, effort from DIH	Access to networks, clusters, associations and federations to which CUT-RCDS-CC belongs and/or has contacts with.
Financial charge for customer	Free.
Funding, business plan	Subsidized by the DIH, research projects.
Brokerage, Matchmaking	
Category	Ecosystem service, Community building.
Description of activity, value proposition (what?)	Bring together local and European stakeholders to set up H2020 proposals.
Format (how?)	Online and face-to-face meetings with stakeholders and local communities.
Profile of DIH contact (who?)	Business facilitators, technical experts.
Involvement, effort from DIH	On-demand meetings based on call complexity.

Financial charge for customer	Free.
Funding, business plan	Subsidized by the DIH and co-organizers.
Market research, market awareness, market/technology trends	
Category	Ecosystem service, Strategy development.
Description of activity, value proposition (what?)	Provide access to experts in market analysis, market data held in the network, at the DIH, or at European and international level.
Format (how?)	Conference call, meeting with business developers at the DIH.
Profile of DIH contact (who?)	Contact with experts in diagnostics, application, and technical mentors.
Involvement, effort from DIH	On demand, depending on the complexity of the task.
Financial charge for customer	Free.
Funding, business plan	Subsidized by the DIH, research projects, joint venture.
Technical consultancy	
Category	Ecosystem service, Strategy development.
Description of activity, value proposition (what?)	Advise on potential uses of robotics and automation.
Format (how?)	Conference calls, bilateral meetings.
Profile of DIH contact (who?)	Technical and application experts.
Involvement, effort from DIH	On demand, depending on the complexity of the task.
Financial charge for customer	Free.
Funding, business plan	Subsidized by the DIH, research projects, joint venture.
Resources and knowledge sharing	
Category	Ecosystem service, Ecosystem learning.
Description of activity, value proposition (what?)	Workshop organization for I&M operators and manufacturers, newsletter distribution, DIH presentations.
Format (how?)	Reference information provided through the DIH, newsletters and RIMA network web portal. Newsletters, workshops, and presentations organized by the DIH or in cooperation with partners.
Profile of DIH contact (who?)	Technical and application mentors.
Involvement, effort from DIH	Technical and application mentors.
Financial charge for customer	Free.
Funding, business plan	Subsidized by the DIH, joint ventures, research projects.
DIH network promotion, Promotion of the offer	
Category	Ecosystem service, Representation, promotion.
Description of activity, value proposition (what?)	Promotion of I&M robotics and relevant organizations, advice to industry partners with regards to marketing plans for robotics in I&M, discussion of marketing support and I&M promotion.

Format (how?)	Dissemination activities, face-to-face meetings, social media, and other activities according to customer interests (e.g., suppliers, end users).
Profile of DIH contact (who?)	Business and technical experts.
Involvement, effort from DIH	Involvement of the technical and business experts.
Financial charge for customer	Free.
Funding, business plan	Subsidized by the DIH, joint ventures, research projects.

2. Business services

Business development	
Category	Incubator/accelerator support.
Description of activity, value proposition (what?)	Advice to industry about IPR management, proposal writing with startups.
Format (how?)	Meetings, conference calls.
Profile of DIH contact (who?)	Business experts.
Involvement, effort from DIH	Meetings with experts in business development.
Financial charge for customer	Free.
Funding, business plan	Subsidized by the DIH, licensing, research projects.
Connect to funding sources	
Category	Access to financing.
Description of activity, value proposition (what?)	Advice to industry for networking with venture capital, funding opportunities, mediating between industry and venture capital.
Format (how?)	Meetings, conference calls, proposal writing.
Profile of DIH contact (who?)	Business developer, technical expert.
Involvement, effort from DIH	Meetings with experts in funding of research, development, and innovation in robotics. Proposal writing.
Financial charge for customer	Free.
Funding, business plan	Subsidized by the DIH, research projects.
Hands-on training	
Category	Skills and education.
Description of activity, value proposition (what?)	Educational and training courses in I&M robotics, STEM education classes for high-school students, underwater I&M seminars and presentations to technology developers, tours of DIH labs and presentations to researchers from other institutions.
Format (how?)	Courses given by CUT personnel. On demand and regular.
Profile of DIH contact (who?)	Technical experts.
Involvement, effort from DIH	1-2 day events and shorter events, supply of training reference material.

Financial charge for customer	Free.
Funding, business plan	Covered by research projects (end-user training), covered by the university (undergraduate/postgraduate courses), subsidized by the DIH (high-school students).
Research, development, and innovation opportunities	
Category	Project development, access to finance.
Description of activity, value proposition (what?)	Advise about technological research, development, and innovation activities as part of proposal writing and research projects. Elaboration of an action plan for RDI with milestones, deliverables, and agreement on the respective roles of stakeholders and on the final delivery.
Format (how?)	Meetings, conference calls.
Profile of DIH contact (who?)	Business developer, technical expert.
Involvement, effort from DIH	Meetings with experts in research, development, and innovation in robotics.
Financial charge for customer	Free.
Funding, business plan	Subsidized by the DIH.
Hosting spin-off and start-up companies	
Category	Offering housing.
Description of activity, value proposition (what?)	Housing offer for the robotics I&M activities of new organizations.
Format (how?)	On demand.
Profile of DIH contact (who?)	Business facilitator.
Involvement, effort from DIH	Office space, workspace, located in DIH premises, internet infrastructure, lab facilities.
Financial charge for customer	Yes.
Funding, business plan	Subsidized by the DIH, joint venture.

3. **Technology services**

Research and development collaboration	
Category	Strategic RDI.
Description of activity, value proposition (what?)	Offering long-term support to collaborating organizations from the private and public sectors through advice and development.
Format (how?)	Collaborative research, development, and innovation.
Profile of DIH contact (who?)	Project manager, technical experts.
Involvement, effort from DIH	Support from experts in the submission of European or national proposals. Long-term participation of several experts from the DIH for technical research and development in robotics.
Financial charge for customer	Yes.
Funding, business plan	Contract research, joint ventures, research projects.

Innovation projects: technology demonstrators, proof of concepts	
Category	Contract research.
Description of activity, value proposition (what?)	Short-term contracts for inspection. Services provided to public sector organizations, research proposals.
Format (how?)	Bilateral or collaborative projects.
Profile of DIH contact (who?)	Project manager, technical experts.
Involvement, effort from DIH	Support from experts in the submission of European or national proposals. Participation of several experts from the DIH for technical research and development in robotics.
Financial charge for customer	Yes.
Funding, business plan	Contract research, joint ventures, research projects.
Proof of concept, prototyping, commercialization/industrialization, small series production	
Category	technical support.
Description of activity, value proposition (what?)	Demonstration in real environment or in realistic conditions. Adaptation of existing technologies to specific use cases.
Format (how?)	Bilateral or collaborative projects.
Profile of DIH contact (who?)	Technical experts.
Involvement, effort from DIH	Support from experts in the submission of European or national proposals. Participation of several experts from the DIH for development and innovation.
Financial charge for customer	Yes.
Funding, business plan	Contract research, joint ventures, research projects.
Lab facilities, equipment for lease/use (sensors, robots, robot tooling, etc.), rapid prototyping	
Category	Provision of technological infrastructure.
Description of activity, value proposition (what?)	Demonstration with equipment held in the lab. Realization of interfaces with DIH equipment. Provision of test benches (for actuation, perception, etc.)
Format (how?)	Conducting experiments with the assistance of CUT personnel.
Profile of DIH contact (who?)	Technical experts.
Involvement, effort from DIH	Support from experts in the submission of European or national proposals.
Financial charge for customer	Yes.
Funding, business plan	Contract research, joint ventures, research projects.

Certification, product demonstration & testing, comparison of products	
Category	Testing and validation.
Description of activity, value proposition (what?)	System testing, demonstration and qualification (mock-up, simulator, mock-up and simulator). Definition of testing, demonstration and qualification KPIs. Run testing, demonstration and qualification scenarios based on the defined KPIs. Demonstration to end-users.
Format (how?)	Conducting experiments with the assistance of CUT personnel following evaluation protocols. Provision of metrics and tools for benchmarking the utility, usability, effectiveness.
Profile of DIH contact (who?)	Technical experts.
Involvement, effort from DIH	Support from experts in the submission of European or national proposals. Participation of several experts from the DIH for maturation.
Financial charge for customer	Yes.
Funding, business plan	Contract research, joint ventures, research projects.

The Danish Robotics DIH

The Danish Robotics DIH comprises DTI, DIRA and Odense Robotics.

BREAKDOWN OF SERVICES

1. Ecosystem services

Awareness creation	
Category	Ecosystem service, Community building.
Description of activity, value proposition (what?)	Robotics awareness presentations for decision-makers and suppliers.
Format (how?)	Webinar/workshops/conferences where the DIH is invited to speak.
Profile of DIH contact (who?)	Senior technical specialists, business developers.
Involvement, effort from DIH	Presentation by a technical specialist and/or business developer.
Financial charge for customer	Free.
Funding, business plan	Subsidized by the DIH.

Dissemination	
Category	Ecosystem service, Community building.
Description of activity, value proposition (what?)	Communication of the services offered by the DIH. Dissemination of results and key facts in local language and/or in English.
Format (how?)	Information distributed through e.g., web sites, social media, newsletters, email.
Profile of DIH contact (who?)	Communication team, technical specialists.
Involvement, effort from DIH	Involvement of the communication teams depends on the format (video, article, newsletter etc.) and extent of the communication effort.
Financial charge for customer	Free.
Funding, business plan	Subsidized by the DIH.

Ecosystem building	
Category	Ecosystem service, Community building.
Description of activity, value proposition (what?)	Putting people in touch with professionals, sectorial associations certification bodies, public and private investors in Denmark and in Europe.
Format (how?)	Reference information provided through the DIH web portal and via DIH contact points.
Profile of DIH contact (who?)	Business developers, technical specialists.
Involvement, effort from DIH	Access to networks, clusters, associations, and federations to which the organizations of the DIH belong and/or has contacts with.
Financial charge for customer	Free.
Funding, business plan	Subsidized by the DIH. Potential economic upside for DIH in generating projects which involve the DIH.

Brokerage, Matchmaking	
Category	Ecosystem service, Community building, Ecosystem learning.
Description of activity, value proposition (what?)	Organization of meetings and workshops between players in the value chain to bring demand and offer sides together to share ideas and analyse opportunities for collaboration.
Format (how?)	Full day / half day event at a central location (e.g., an airport) or online with presentations from stakeholders, networking lunch (if not digital), dialogue B2B, tour of the DIH if visit on site.
Profile of DIH contact (who?)	Business developers, technical specialists.
Involvement, effort from DIH	One-day event with participation of 2-4 specialists from the DIH, communication material, coordination of event, practicalities on the day (providing lunch, coffee etc.).
Financial charge for customer	Free / participation fee.
Funding, business plan	Subsidized by the DIH, co-financed by the national cluster for automation, robotics, and drones (Odense Robotics) or similar.
Market research, market awareness, market/technology trends	
Category	Ecosystem service, Strategy development.
Description of activity, value proposition (what?)	Provide access to experts in market analysis, market data held in the network, at the DIH, and at European and international level.
Format (how?)	Conference call, meeting with business developers at the DIH.
Profile of DIH contact (who?)	Contact to experts in market analysis.
Involvement, effort from DIH	One hour with experts.
Financial charge for customer	Free or paid.
Funding, business plan	Subsidized by the DIH.

Mentoring by technical and application experts	
Category	Ecosystem service, Strategy development.
Description of activity, value proposition (what?)	Advise on possible uses of robotics and automation.
Format (how?)	Conference calls, bilateral meetings, meetings, or workshops at customer's premises or DIH.
Profile of DIH contact (who?)	Technical and application specialists, business developers.
Involvement, effort from DIH	1-4 hours with 1-2 specialists.
Financial charge for customer	Free (typically 1-2 hours, online) or paid if more effort is required.
Funding, business plan	Subsidized by the DIH (as part of investigating project opportunities) or cost per hour for larger efforts. In some cases, a mix of public and private funding.

Resources and knowledge sharing	
Category	Ecosystem service, Ecosystem learning.
Description of activity, value proposition (what?)	Informing about and organizing access to resources, databases, seminars, workshops, conferences, field trips, newsletters, and other information of interest.
Format (how?)	Reference information provided through the DIH, newsletters and RIMA network web portal. Conferences, field trips, workshops organized by the DIH or in cooperation with partners.
Profile of DIH contact (who?)	Communication team.
Involvement, effort from DIH	Involvement of the communication teams of the organizations involved in the DIH.
Financial charge for customer	Free (information sharing, some conferences) or paid (e.g., conference fee)
Funding, business plan	Subsidized by the DIH. Mixed public and private funding.

DIH network promotion, Promotion of the offer	
Category	Ecosystem service, Ecosystem learning.
Description of activity, value proposition (what?)	Promote both the DIH and its members.
Format (how?)	Webpages, newsletters. participation in roadshows, exhibitions, dissemination, and other activities according to customer interests (e.g., suppliers, end users).
Profile of DIH contact (who?)	Communication, business developers and technical specialists.
Involvement, effort from DIH	Involvement of the communication teams, business developers and technical experts.
Financial charge for customer	Free or paid.
Funding, business plan	Subsidized by the DIH. Mix of public and private funding.

2. Business services

Business development[1]	
Category	Incubator/accelerator support.
Description of activity, value proposition (what?)	Support for business development, facilitating the creation of start-ups, protecting research results, advising about IP management, carrying out strategic analyses, finding scientific and technical information, business intelligence gathering. Setting up industrial and collaborative contracts.
Format (how?)	Meetings, conference calls, project management.

1 regarding DTI technologies

Profile of DIH contact (who?)	Experts in business development, in IPR management.
Involvement, effort from DIH	Meetings with experts in business development.
Financial charge for customer	Free for DTI.
Funding, business plan	Mixed public funding and internal investment in DIH.
Connect to funding sources	
Category	Access to financing, Incubator/accelerator support.
Description of activity, value proposition (what?)	Connect businesses and public organizations to relevant funding opportunities including tax refund mechanisms. Lead or contribute to writing applications for the funding opportunities. Business development as part of project definition.
Format (how?)	Meetings, conference calls, proposal drafts.
Profile of DIH contact (who?)	Business developer, technical specialists.
Involvement, effort from DIH	Meetings with experts in funding of research, development and innovation in robotics. Proposal writing.
Financial charge for customer	Free (sales activity, or as limited support offer) or paid (DIH involvement only in proposal/funding investigating phase)
Funding, business plan	Subsidized by the DIH when as part of sales activity or as limited support offer (e.g., a 1-hour information meeting). Private funding when DIH involvement is only in sales phase and not in a potential follow-up project.
Hands-on training	
Category	Skills and education.
Description of activity, value proposition (what?)	Training courses for technical personnel (end-users, operators etc.)
Format (how?)	Courses given by DIH specialists. On demand and regular.
Profile of DIH contact (who?)	Technical specialists.
Involvement, effort from DIH	1-2 day training courses and shorter events, supply of training reference material.
Financial charge for customer	Yes.
Funding, business plan	Flat fee per seminar/training course.

Research, development, and innovation opportunities	
Category	Project development, access to finance.
Description of activity, value proposition (what?)	Advise about technological RDI. Elaboration of an action plan for RDI with milestones, deliverables, agreement on the respective role of stakeholders and on the final delivery. Basis for initiating, e.g., bilateral projects between industry and RTOs in the DIH, or as basis for proposal writing in response to funding opportunities.

Format (how?)	Meetings, conference calls.
Profile of DIH contact (who?)	Business developer, technical specialists.
Involvement, effort from DIH	Meetings with experts in RDI in robotics.
Financial charge for customer	Free if the activity can lead to project for the DIH or paid else.
Funding, business plan	Subsidized by the DIH. Private finding.
Hosting spin-off and start-up companies	
Category	Offering housing.
Description of activity, value proposition (what?)	Hosting and mentoring of robotics start-ups.
Format (how?)	On demand.
Profile of DIH contact (who?)	Business facilitator.
Involvement, effort from DIH	Office space, workspace, located in DIH premises, internet infrastructure, lab facilities, expert support, etc.
Financial charge for customer	Free.
Funding, business plan	Subsidized by DTI with funding from private investors.

3. Technology services

Research and development collaboration	
Category	Strategic RDI.
Description of activity, value proposition (what?)	Research aiming to reach TRL 5, proof-of-concept, specifications, research and development, validation in laboratory.
Format (how?)	Collaborative research with several partners in DK, EU and/or outside the EU for 2 years or more.
Profile of DIH contact (who?)	Project manager, technical specialists.
Involvement, effort from DIH	Support from experts in the submission of European or national proposals. Long term participation (over 2 years) of several experts from the DIH for technical RDI in robotics.
Financial charge for customer	Free or paid, sometimes with in-kind contributions from industry.
Funding, business plan	Funded by European or national research programmes, mix of public and private funding, or solely private funding (rare cases).
Innovation projects: technology demonstrators, proof of concepts	
Category	Contract research.
Description of activity, value proposition (what?)	Research and development aiming to reach TRL 7, application of existing technologies, validation in laboratory.
Format (how?)	Bilateral or collaborative projects from one to a few years, using any robotic technologies.
Profile of DIH contact (who?)	Project manager, technical specialists, business developers.

Involvement, effort from DIH	Support from experts in the submission of European or national proposals. Participation of several experts from the DIH for technical RDI in robotics.
Financial charge for customer	Yes.
Funding, business plan	Funded by bilateral contracts, European or national innovation programmes, private investors.
Proof of concept, prototyping, commercialization/industrialization, small series production	
Category	Technical support on scale-up.
Description of activity, value proposition (what?)	Demonstration in real environment or in realistic conditions. Adaptation of existing technologies to specific use case.
Format (how?)	Bilateral projects from one to several months, using anyone's technologies.
Profile of DIH contact (who?)	Technical specialists.
Involvement, effort from DIH	Support from experts in the submission of European or national proposals. Participation of several experts from the DIH for development and innovation.
Financial charge for customer	Yes.
Funding, business plan	Funded by bilateral contracts, European or national innovation programmes, investors etc.
Lab facilities, equipment for lease/use (sensors, robots, robot tooling, etc.), rapid prototyping	
Category	Provision of technological infrastructure.
Description of activity, value proposition (what?)	Demonstration with equipment held in the laboratories at the DIH or deployment of sensors etc. at customer site. Realization of interfaces with DIH equipment. Provision of test benches (for actuation, perception, etc.). Provision of large-scale field laboratories (e.g., test area for drones at HCA Airport).
Format (how?)	Conducting experiments with the assistance of DIH personnel.
Profile of DIH contact (who?)	Technical specialists.
Involvement, effort from DIH	Support from experts in the submission of European or national proposals. Participation of several experts from the DIH for maturation.
Financial charge for customer	Financial charge based on daily rate and depending on the personnel involved, equipment used, and adaptation needed.
Funding, business plan	Funded by bilateral contracts, European or national innovation programmes, investors.

DIH Paris-Saclay

The DIH Paris-Saclay or DIGIHALL comprises CEA and Systematic.

BREAKDOWN OF SERVICES

1. Ecosystem services

Awareness creation	
Category	Ecosystem service, Community building.
Description of activity, value proposition (what?)	Robotics awareness course for decision-makers.
Format (how?)	Workshop with hands-on approach, webinar.
Profile of DIH contact (who?)	Technical experts.
Involvement, effort from DIH	One half-day event with senior experts from the DIH providing the course.
Financial charge for customer	Free.
Funding, business plan	Subsidized by the DIH.
Dissemination	
Category	Ecosystem service, Community building.
Description of activity, value proposition (what?)	Communication of the services offered by the DIH. Dissemination of results and key facts in local language and in English.
Format (how?)	Information distributed through e.g., web sites, social media, newsletters, email.
Profile of DIH contact (who?)	Experts in communication.
Involvement, effort from DIH	Involvement of the communication teams belonging to the organizations involved in the DIH at CEA and SYSTEMATIC.
Financial charge for customer	Free.
Funding, business plan	Subsidized by the DIH.

Ecosystem building	
Category	Ecosystem service, Community building.
Description of activity, value proposition (what?)	Putting people in touch with professional and sectorial associations, certification bodies, public and private investors in the local region and throughout Europe.
Format (how?)	Reference information provided through the DIH web portal. Possible contact with experts.
Profile of DIH contact (who?)	Business facilitators, technical experts.
Involvement, effort from DIH	Access to networks, clusters, associations, and federations to which the organizations of the DIH belong.
Financial charge for customer	Free.
Funding, business plan	Subsidized by the DIH.

Brokerage, Matchmaking	
Category	Ecosystem service, Community building.
Description of activity, value proposition (what?)	Organization of meetings between players in the value chain to bring demand and offer sides together to share ideas and analyse opportunities.
Format (how?)	One event per year at the DIH or online with presentations from stakeholders, visit to the DIH site, networking lunch when on site and dialogue B2B space.
Profile of DIH contact (who?)	Business facilitators, technical experts.
Involvement, effort from DIH	One-day event with participation of 3 to 4 experts from the DIH, supply of communication material.
Financial charge for customer	Free.
Funding, business plan	Subsidized by the DIH.

Market research, market awareness, market/technology trends	
Category	Ecosystem service, Strategy development.
Description of activity, value proposition (what?)	Provide access to experts in market analysis, market data held in the network, at the DIH, at European and international level.
Format (how?)	Conference call, meeting with business developers at the DIH.
Profile of DIH contact (who?)	Access to experts in market analysis.
Involvement, effort from DIH	One hour with experts.
Financial charge for customer	Yes.
Funding, business plan	Subsidized by the DIH.
Technical consultancy	
Category	Ecosystem service, Strategy development.
Description of activity, value proposition (what?)	Advise on pontential uses of robotics and automation.
Format (how?)	Conference calls, bilateral meetings.
Profile of DIH contact (who?)	Technical and application experts.
Involvement, effort from DIH	One hour with experts.
Financial charge for customer	Free.
Funding, business plan	Subsidized by the DIH.
Resources and knowledge sharing	
Category	Ecosystem service, Ecosystem learning
Description of activity, value proposition (what?)	Informing about and organizing access to resources, databases, seminars, workshops, conferences, field trips, newsletters, and other information of interest.
Format (how?)	Reference information provided through the DIH and RIMA network web portal. Possible contact with experts for advanced advice.
Profile of DIH contact (who?)	Systematic cluster operational team (communication, innovation)

Involvement, effort from DIH	Involvement of the communication teams at CEA and SYSTEMATIC.
Financial charge for customer	Free for Systematic members.
Funding, business plan	Mixed public and private funding.
DIH network promotion, Promotion of the offer	
Category	Ecosystem service, Ecosystem learning.
Description of activity, value proposition (what?)	Promote both the DIH and its members.
Format (how?)	Organization of roadshows, exhibitions, dissemination, and other activities according to customer interests (e.g., suppliers, end users).
Profile of DIH contact (who?)	Communication, business, and technical experts.
Involvement, effort from DIH	Involvement of the communication teams belonging to CEA and SYSTEMATIC.
Financial charge for customer	Free.
Funding, business plan	Subsidized by the DIH.

2. **Business services**

Business development[2]	
Category	Incubator/accelerator support.
Description of activity, value proposition (what?)	Dedicated to spin-off and start-up companies promoting CEA technologies. Support for business creation, facilitating the creation of start-ups, protecting research results, advising about IP management, carrying out strategic analyses, finding scientific and technical information and business intelligence, setting up industrial and collaborative contracts.
Format (how?)	Meetings, conference calls.
Profile of DIH contact (who?)	Experts in business creation, in IPR management.
Involvement, effort from DIH	Meetings with experts in business development.
Financial charge for customer	Free for Systematic members.
Funding, business plan	N/A
Funding opportunities investigation	
Category	Incubator/accelerator support.
Description of activity, value proposition (what?)	Dedicated to organizations, especially spin-offs and start-ups interested in promoting CEA technologies or willing to do collaborative research with CEA. Advice on financial support according to maturity level. Presentation of opportunities offered by European research and innovation programmes, regional sources of funding, tax refund mechanism and private investors. Support for business creation, facilitating the creation of start-ups, protecting

2 regarding CEA technologies

	research results, advising about IP management, carrying out strategic analyses, finding scientific and technical information, business intelligence, setting up industrial and collaborative contracts.
Format (how?)	Meetings, conference calls.
Profile of DIH contact (who?)	Business developer, technical expert, expert in IPR management.
Involvement, effort from DIH	Meetings with experts in funding of RDI in robotics.
Financial charge for customer	Free for Systematic members.
Funding, business plan	Subsidized by the DIH.
Hands-on training	
Category	Skills and education.
Description of activity, value proposition (what?)	Training of operators, raising awareness about teleoperation, and robotics.
Format (how?)	Courses and hands-on training on equipment available at the DIH, on demand.
Profile of DIH contact (who?)	Technical experts.
Involvement, effort from DIH	One day with participation of experts, supply of training reference material.
Financial charge for customer	Free for Systematic members.
Funding, business plan	Subsidized by the DIH.
Research, development and innovation opportunities	
Category	Project development.
Description of activity, value proposition (what?)	Dedicated to organizations interested in promoting CEA technologies or willing to do collaborative research with CEA. Advise about technological research, development and innovation activity. Elaboration of an action plan for RDI with milestones, deliverables, agreement on the respective role of stakeholders and on the final delivery.
Format (how?)	Meetings, conference calls.
Profile of DIH contact (who?)	Business developer, technical experts.
Involvement, effort from DIH	Meetings with experts in RDI in robotics.
Financial charge for customer	Free.
Funding, business plan	Subsidized by the DIH.
Hosting spin-off and start-up companies	
Category	Offering housing.
Description of activity, value proposition (what?)	Accompany and host follow-up of spin-off or start-up companies willing to promote CEA technologies, patents, licenses.
Format (how?)	On demand.
Profile of DIH contact (who?)	Business facilitator.
Involvement, effort from DIH	Office space, workspace, located in DIH premises at CEA Nano-Innov, internet infrastructure.
Financial charge for customer	Yes.
Funding, business plan	Rates according to needs.

3. Technology services

Research and development collaboration	
Category	Strategic RDI.
Description of activity, value proposition (what?)	Research aiming to reach TRL 5, removal of doubts, specifications, research and development, validation in laboratory.
Format (how?)	Collaborative research with several partners in EU lasting 2 years or more.
Profile of DIH contact (who?)	Project manager, technical experts.
Involvement, effort from DIH	Support from experts in the submission of European or national proposals. Long term participation (over 2 years) of several experts from the DIH for technical research and development in robotics.
Financial charge for customer	Free.
Funding, business plan	Funded by European or national research programmes.
Innovation projects: technology demonstrators, proof of concepts	
Category	Contract research.
Description of activity, value proposition (what?)	Research and development aiming to reach TRL 6, application of existing technologies, validation in laboratory.
Format (how?)	Bilateral projects from one to a few years.
Profile of DIH contact (who?)	Project manager, technical experts.
Involvement, effort from DIH	Support from experts in the submission of European or national proposals. Long term participation (over 2 years) of several experts from the DIH for technical research and development in robotics.
Financial charge for customer	Yes.
Funding, business plan	Funded by bilateral contracts, European or national innovation programmes, private investors.
Proof of concept, prototyping, commercialisation/industrialisation, small series production	
Category	Technical support on scale-up.
Description of activity, value proposition (what?)	Demonstration in real environment or in realistic conditions. Adaptation of existing technologies to specific use case.
Format (how?)	Bilateral projects from one to several months.
Profile of DIH contact (who?)	Technical experts.
Involvement, effort from DIH	Support from experts in the submission of European or national proposals. Participation of several experts from the DIH for development and innovation.
Financial charge for customer	Yes.
Funding, business plan	Funded by bilateral contracts, European or national innovation programmes, investors.

Lab facilities, equipment for lease/use (sensors, robots, robot tooling, etc.), rapid prototyping	
Category	Provision of technological infrastructure.
Description of activity, value proposition (what?)	Demonstration with equipment held in the lab. Realization of interfaces with DIH equipment. Provision of test benches (for actuation, perception, etc.)
Format (how?)	Conducting experiments with the assistance of CEA personnel.
Profile of DIH contact (who?)	Technical experts.
Involvement, effort from DIH	Support from experts in the submission of European or national proposals. Participation of several experts from the DIH for maturation.
Financial charge for customer	Yes. The financial charge is based on daily rate and depending on the personnel involved, equipment used at the DIH and adaptation needed.
Funding, business plan	Funded by bilateral contracts, European or national innovation programmes, investors.
Certification, product demonstration & testing, comparison of products	
Category	Testing and validation.
Description of activity, value proposition (what?)	For testing and benchmarking, CEA holds and can offer access to a set of dedicated infrastructures for running assessments in realistic conditions and in particular for applications in nuclear sector. CEA can also organize access to infrastructure in other application domains.
Format (how?)	Conducting experiments with the assistance of CEA personnel following evaluation protocols. Provision of metrics and tools for benchmarking utility, usability, effectiveness.
Profile of DIH contact (who?)	Technical experts. For certification issues, CEA will provide contacts to experts of CETIM[3] the Technical Centre for Mechanical Industry.
Involvement, effort from DIH	Support from experts in the submission of European or national proposals. Participation of several experts from the DIH for maturation.
Financial charge for customer	Yes.
Funding, business plan	Funded by bilateral contracts, European or national innovation programmes, investors.

3 https://m.cetim.fr/#

The German Aerospace Centre (DLR)

BREAKDOWN OF SERVICES

1. Ecosystem services

Awareness creation	
Category	Ecosystem service, Community building.
Description of activity, value proposition (what?)	Organizing events promoting robotics technologies and applications.
Format (how?)	Events, meetings, webinars.
Profile of DIH contact (who?)	Developers, technical experts, business developers.
Involvement, effort from DIH	Half or one day events.
Financial charge for customer	Free.
Funding, business plan	Operational costs of DIH.
Technical consultancy	
Category	Ecosystem service, Strategy development.
Description of activity, value proposition (what?)	Consultancy on applications for robotic technologies and systems.
Format (how?)	Meetings, phone calls, webinars.
Profile of DIH contact (who?)	Technical and application experts.
Involvement, effort from DIH	Short meetings with clients.
Financial charge for customer	Free.
Funding, business plan	Subsidized by the DIH.

2. Business services

Funding opportunities investigation	
Category	Incubator/accelerator support.
Description of activity, value proposition (what?)	Advising about funding opportunities, support for the applications.
Format (how?)	Meetings.
Profile of DIH contact (who?)	Business developers.
Involvement, effort from DIH	DIH business developers, senior project leaders.
Financial charge for customer	Free.
Funding, business plan	Subsidized by the DIH.
Research, development and innovation opportunities	
Category	Project development.
Description of activity, value proposition (what?)	Support for planning of R&D projects.
Format (how?)	Meetings, calls.
Profile of DIH contact (who?)	Business developer, project leader, technical expert.
Involvement, effort from DIH	Meetings with researchers and project leader.
Financial charge for customer	Free.
Funding, business plan	Subsidized by the DIH.

3. Technology services

Research and development collaboration	
Category	Strategic RDI.
Description of activity, value proposition (what?)	Validation of new technologies and systems at lower TRLs, consultancy, joint projects.
Format (how?)	Collaborative research with partners from industry and SMEs.
Profile of DIH contact (who?)	Project manager, technical experts.
Involvement, effort from DIH	Support for preparation and submission of European or national proposals. Conduction of joint projects (usually 1 - 3 years) with experts from the DIH.
Financial charge for customer	Free.
Funding, business plan	Funded by European or national research programmes.
Innovation projects: technology demonstrators, proof of concepts	
Category	Contract research.
Description of activity, value proposition (what?)	Validation of new technologies and systems at TRL 6, consultancy, joint projects.
Format (how?)	Short term projects with partners (from a few months to a few years)
Profile of DIH contact (who?)	Senior researchers, technical experts.
Involvement, effort from DIH	Support from experts to develop concepts and technical solutions for customer-defined problems.
Financial charge for customer	Yes.
Funding, business plan	Funded through contracts with customer, national and EU innovation programmes.

Lab facilities, equipment for lease/use (sensors, robots, robot tooling, etc.), rapid prototyping	
Category	Provision of technological infrastructure.
Description of activity, value proposition (what?)	Experiments and testing of customer technologies on aerial systems of the Institute of Robotics and Mechatronics.
Format (how?)	Conducting experiments with the assistance of DIH personnel, joint short-term projects
Profile of DIH contact (who?)	Technical experts, senior researchers.
Involvement, effort from DIH	Planning, preparation and execution of experiments or short-term projects with customers.
Financial charge for customer	Yes.
Funding, business plan	Funded through contracts with customer, national and EU innovation programmes.

The German Research Center for Artificial Intelligence (DFKI)

BREAKDOWN OF SERVICES

1. Ecosystem services

Awareness creation	
Category	Ecosystem service, Community building.
Description of activity, value proposition (what?)	Introduction to robotics and AI for industrial decision-makers.
Format (how?)	Seminars and workshops with hands-on approach, both physical and online (webinars)
Profile of DIH contact (who?)	Technical experts, business developers.
Involvement, effort from DIH	Short (one half day) to long (3 days) events involving senior and junior experts from DFKI.
Financial charge for customer	Yes: fee based on content and duration of course.
Funding, business plan	Paid by customer or through relevant projects.
Dissemination	
Category	Ecosystem service, Community building.
Description of activity, value proposition (what?)	Communication about the services offered by the DIH. Dissemination of results and key facts in German and in English.
Format (how?)	Information distributed through e.g., web sites, social media, newsletters, email.
Profile of DIH contact (who?)	Experts in communication.
Involvement, effort from DIH	Involvement of the DFKI communications team.
Financial charge for customer	Free.
Funding, business plan	Subsidized by the DIH as part of business development.

Technical consultancy	
Category	Ecosystem service, Strategy development.
Description of activity, value proposition (what?)	Advise on the potential uses of robotics, automation, and artificial intelligence.
Format (how?)	Bilateral meetings and calls.
Profile of DIH contact (who?)	Technical and application experts.
Involvement, effort from DIH	Short meetings with clients.
Financial charge for customer	Free.
Funding, business plan	Subsidized by the DIH as part of business development.
DIH network promotion, Promotion of the offer	
Category	Ecosystem service, Ecosystem learning.
Description of activity, value proposition (what?)	Promote both the DIH and its services.
Format (how?)	Presentations at dissemination events, and on online media, social media.
Profile of DIH contact (who?)	Experts from the communications department, business developers.

Involvement, effort from DIH	Participation in events, content maintenance of online media and social media.
Financial charge for customer	Free.
Funding, business plan	Subsidized by the DIH.

2. Business services

Funding opportunities investigation	
Category	Incubator/accelerator support.
Description of activity, value proposition (what?)	Investigation of funding opportunities and advice to customers about their options, based on their needs.
Format (how?)	Meetings.
Profile of DIH contact (who?)	Business developers.
Involvement, effort from DIH	Participation of DIH business developers in information events, research on the web, evaluation of newsletters etc.; meetings and bi-lateral calls with clients.
Financial charge for customer	Free.
Funding, business plan	Subsidized by the DIH.
Hands-on training	
Category	Skills and education.
Description of activity, value proposition (what?)	Training of operators (e.g. in underwater robots), practical courses in robotics.
Format (how?)	Courses and hands-on training using equipment and facilities of DIH. On demand.
Profile of DIH contact (who?)	Technical expert, researcher.
Involvement, effort from DIH	Provision of equipment and facilities, technical experts, and researchers for training.
Financial charge for customer	Free or paid, depending on client status.
Funding, business plan	Rates according to course content and scope.
Research, development, and innovation opportunities	
Category	Project development.
Description of activity, value proposition (what?)	Dedicated to organizations interested in promoting DFKI technologies or willing to do collaborative research with DFKI. Advising on technological research, development and innovation activity. Elaboration of an action plan for RDI with milestones, deliverables, agreement on the respective role of stakeholders and on the final delivery.
Format (how?)	Meetings, conference calls.
Profile of DIH contact (who?)	Business developer, technical expert.
Involvement, effort from DIH	Meetings with experts in research, development and innovation in robotics.
Financial charge for customer	Free.
Funding, business plan	Subsidized by the DIH.

3. Technology services

Research and development collaboration	
Category	Strategic RDI.
Description of activity, value proposition (what?)	Research aiming to reach TRL 5, removal of doubts, specifications, research and development, validation in laboratory.
Format (how?)	Collaborative research with partners from industry and SMEs, in Germany and in the EU, usually 3 years, sometimes less.
Profile of DIH contact (who?)	Project manager, technical experts.
Involvement, effort from DIH	Support from experts in the writing and submission of European or national proposals. Long term participation (usually 3 years) of several experts from the DIH for technical research and development in robotics.
Financial charge for customer	Free.
Funding, business plan	Funded by European or national research programmes.
Innovation projects: technology demonstrators, proof of concepts	
Category	Contract research.
Description of activity, value proposition (what?)	Research and development aiming to reach TRL 6, application of existing technologies, validation in laboratory.
Format (how?)	Bilateral or multi-partner projects with shorter terms (from a few months to a few years)
Profile of DIH contact (who?)	Technical experts.
Involvement, effort from DIH	Support from experts to develop concepts and technical solutions for customer-defined problems.
Financial charge for customer	Yes.
Funding, business plan	Funded through customer based on bilateral contracts, private investors, national and EU innovation programmes.
Proof of concept, prototyping, commercialization/industrialization, small series production	
Category	Technical support on scale-up.
Description of activity, value proposition (what?)	Demonstration in real environment or in realistic conditions. Adaptation of existing technologies to the specific use case.
Format (how?)	Bilateral projects from one to several months.
Profile of DIH contact (who?)	Technical experts.
Involvement, effort from DIH	Support from experts to develop concepts and technical solutions for customer-defined problems.
Financial charge for customer	Yes.
Funding, business plan	Funded through customer based on bilateral contracts, private investors, national and EU innovation programmes.

Lab facilities, equipment for lease/use (sensors, robots, robot tooling, etc.), rapid prototyping	
Category	Provision of technological infrastructure.
Description of activity, value proposition (what?)	Testing of customer equipment in DFKI lab and testing facilities.
Format (how?)	Conducting experiments with the assistance of DFKI personnel.
Profile of DIH contact (who?)	Lab technicians, technical experts.
Involvement, effort from DIH	Preparation and maintenance of lab facilities; supervision and support of client activities by DFKI technicians and experts.
Financial charge for customer	Yes: flat fee based on cost table and complexity of experiment.
Funding, business plan	Client-funded based on facility leasing contract, sometimes covered by European or national innovation programmes.

Certification, product demonstration & testing, comparison of products	
Category	Testing and validation.
Description of activity, value proposition (what?)	Evaluation of the pressure compliance of technical components in our pressure lab.
Format (how?)	Experiments by DFKI experts following standard evaluation protocols.
Profile of DIH contact (who?)	Technical experts.
Involvement, effort from DIH	Preparation, maintenance and provision of pressure lab; support of experiments by DIH experts.
Financial charge for customer	Yes: flat-fee based on cost table and complexity of experiment.
Funding, business plan	Client-funded based on facility leasing contract, sometimes covered by European or national innovation programmes.

PIAP HUB

The industrial research institute for automation and measurements PIAP is part of the Łukasiewicz research network.

BREAKDOWN OF SERVICES

1. Ecosystem services

Awareness creation	
Category	Ecosystem service, Community building.
Description of activity, value proposition (what?)	Robotics awareness presentation for decision-makers and suppliers.
Format (how?)	Webinar/workshops/conferences where the DIH is invited to speak.
Profile of DIH contact (who?)	Senior technical experts.
Involvement, effort from DIH	Presentation by one technical expert.
Financial charge for customer	Free.
Funding, business plan	Subsidized by the DIH.
Dissemination	
Category	Ecosystem service, Community building.
Description of activity, value proposition (what?)	Communication of the DIH service offer. Dissemination of results and key facts in the local language and in English.
Format (how?)	Information distributed through e.g. web sites, social media, newsletters, email.
Profile of DIH contact (who?)	Experts in communication, technical experts.
Involvement, effort from DIH	Involvement of the communication teams belonging to the organizations involved in the DIH.
Financial charge for customer	Free.
Funding, business plan	Subsidized by the DIH.

Brokerage, Matchmaking	
Category	Ecosystem service, Community building, Ecosystem learning.
Description of activity, value proposition (what?)	Organization of meetings and workshops between players in the value chain to bring demand and offer sides together to share ideas and analyse opportunities for collaboration.
Format (how?)	Minimum one event per year at the PIAP conference center or online with presentations from stakeholders and networking coffee (if not digital).
Profile of DIH contact (who?)	Business facilitators, technical experts.
Involvement, effort from DIH	One day event with participation of 4 to 5 experts from the DIH and stakeholder representatives.
Financial charge for customer	Depending on the event free or partially funded by the participants.
Funding, business plan	Funded by sponsors, participants or public funding, depending on topic.

Mentoring by technical and application experts	
Category	Ecosystem service, Strategy development.
Description of activity, value proposition (what?)	Advise on the potential uses of robotics and automation.
Format (how?)	Conference calls, bilateral meetings, meetings on customer's premises.
Profile of DIH contact (who?)	Technical and application experts.
Involvement, effort from DIH	One hour or more with experts.
Financial charge for customer	Free (typically 1-2 hours, online or on-site)
Funding, business plan	Subsidized by the DIH (as part of investigating project opportunities) or cost per hour for larger events. In some cases a mix of public and private funding was used.
Resources and knowledge sharing	
Category	Ecosystem service, Ecosystem learning.
Description of activity, value proposition (what?)	Informing about and organizing access to resources, databases, seminars, workshops, conferences, field trips, newsletters and other information of interest.
Format (how?)	Reference information provided through the DIH newsletters and RIMA network web portal. Conferences, field trips, workshops organized by the DIH or in cooperation with partners.
Profile of DIH contact (who?)	Communication team. Experts
Involvement, effort from DIH	Involvement of the communication teams of the organizations involved in the DIH at PIAP
Financial charge for customer	Free (information sharing, some conferences) or paid (e.g., conference fee)
Funding, business plan	Mixed public and private funding.

2. **Business services**

Connect to funding sources	
Category	Access to financing, Incubator/accelerator support.
Description of activity, value proposition (what?)	Connect businesses and public organizations to relevant funding opportunities. Lead or contribute to proposal writing in response to funding opportunities. Business development as part of proposal writing. Financial project management support. Support in case of audit.
Format (how?)	Meetings, conference calls, proposal writing.
Profile of DIH contact (who?)	Business developer, technical expert, financial expert.
Involvement, effort from DIH	Meetings with experts in the funding of RDI in robotics. Proposal writing. Project management.

Financial charge for customer	Free (sales activity, or as limited support offer provided by ŁPIAP) or paid (DIH involvement only in proposal/funding investigation phase, core service provided by another entity from the DIH partnership)
Funding, business plan	Subsidized by the DIH when part of a sales activity or as a limited-support offer (e.g., a 1-hour information meeting). Private funding when the DIH is involved only in the sales phase and not in the potential subsequent project.
Hands-on training	
Category	Skills and education.
Description of activity, value proposition (what?)	Training of technical personnel.
Format (how?)	Courses given by DIH personnel on demand.
Profile of DIH contact (who?)	Technical experts.
Involvement, effort from DIH	1-2 day events and shorter events, supply of training reference material.
Financial charge for customer	Yes.
Funding, business plan	Individually negotiated fee depending on the size of the group and scope of training.

3. Technology services

Innovation projects: technology demonstrators, proof of concepts	
Category	Contract research.
Description of activity, value proposition (what?)	Research and development aiming to reach TRL 6, application of existing technologies, validation in laboratory.
Format (how?)	Bilateral or collaborative projects from one to a few years.
Profile of DIH contact (who?)	Project manager, technical experts.
Involvement, effort from DIH	Support from experts in the submission of European or national proposals. Participation of several experts from the DIH for technical research and development in robotics.
Financial charge for customer	Yes.
Funding, business plan	Funded by bilateral contracts, European or national innovation programmes.
Proof of concept, prototyping, commercialization/industrialization, small series production	
Category	Technical support on scale-up.
Description of activity, value proposition (what?)	Demonstration in real environment or in realistic conditions. Adaptation of existing technologies to the specific use case. Technology implementation to TRL 7 -9.
Format (how?)	Bilateral projects from one to several months.
Profile of DIH contact (who?)	Technical experts.

Involvement, effort from DIH	Support from experts in the submission of European or national proposals. Participation of several experts from the DIH for development and innovation.
Financial charge for customer	Yes.
Funding, business plan	Funded by bilateral contracts, European or national innovation programmes, investors.

Lab facilities, equipment for lease/use (sensors, robots, robot tooling, etc.), rapid prototyping	
Category	Provision of technological infrastructure.
Description of activity, value proposition (what?)	Demonstration with equipment held in the laboratories at the DIH or deployment of sensors etc. at customer site. Realization of interfaces with DIH equipment. Provision of test benches (for actuation, perception, etc.). Provision of large-scale field laboratories (e.g., test area for autonomous ground platforms, and other labs described earlier in this document)
Format (how?)	Conducting experiments with the assistance of DIH personnel.
Profile of DIH contact (who?)	Technical experts.
Involvement, effort from DIH	Support from experts in the submission of European or national proposals. Participation of several experts from the DIH for maturation.
Financial charge for customer	Yes.
Funding, business plan	Funded by bilateral contracts, European or national innovation programmes, investors.

Laboratory for Manufacturing Systems & Automation (LMS)

BREAKDOWN OF SERVICES

Different format because of LMS preference.

1. Business Services

- Access to finance – Incubator/accelerator support:

Description of activity, value proposition (what?): Connect businesses and public organizations to relevant funding opportunities. Lead or contribute to proposal writing in response to funding opportunities.

- Project development:

Description of activity, value proposition (what?): Consultancy regarding technological research, development and innovation activities. Elaboration of an action plan for RDI with milestones, deliverables, agreement on the respective role of stakeholders and on the final delivery. Basis for initiating bilateral projects between industry and RTOs in the DIH or proposal writing to secure funding.

- Skills and education:

Description of activity, value proposition (what?): Training of undergraduate / postgraduate students

- Offering housing:

Description of activity, value proposition (what?): Offer space to set up demonstrators and perform proof of concepts and technology validation within the scope of a project.

- General information on business services:

Format (how?): Online or physical meetings, conference calls, proposal writing, lectures and lab courses given by DIH personnel.

Profile of DIH contact (who?): Project managers, technical research engineers and senior research engineers.

Involvement, effort from DIH: Meetings with experts in funding of research, development and innovation in robotics. Two Course on a yearly basis, supply of training reference material.

Funding, business plan: Subsidized by the DIH (part of investigating project opportunities).

Financial charge for customer: Free or part of project investigation.

2. Ecosystem Services

- Ecosystem learning – Community building:

Description of activity, value proposition (what?): Organisation of meetings between industries in the value chain to bring demand and offer sides together to share ideas and analyse opportunities for collaboration. Communication of the key activities of the DIH. Dissemination of results and key facts from European and national projects where LMS is involved. Putting people in touch with professional and sectorial associations, certification bodies, public and private investors in our region and in Europe.

- Strategy development:

Description of activity, value proposition (what?): Advise on potential uses of robotics and automation.

- General information on ecosystem services:

Format (how?): Meetings taking place at LMS premises: tour of the laboratory, discussion and presentations from stakeholders, networking lunch, B2B dialogue. Webinar/workshops/conferences where the DIH is invited to speak; digital information distributed through e.g., web sites, social media, newsletters, email, videos, presentations, videos of live demonstrations.

Profile of DIH contact (who?): Business facilitators, technical experts, project managers and senior research engineers.

Involvement, effort from DIH: Participation of LMS project managers in events; access to networks, clusters, associations and federations to which the organizations of the DIH belong and/or have contact with; supply of communication material.

Funding, business plan: Subsidized by the DIH (as part of investigating project opportunities).

Financial charge for customer: Free or part of project investigation.

3. Technology Services

- Contract Research:

Description of activity, value proposition (what?): Research and development aiming to reach TRL 6, application of existing technologies, validation in lab.

- Provision of technology & infrastructure:

Description of activity, value proposition (what?): Demonstration with equipment held in the laboratories at the DIH or deployment of sensors etc. at customer site. Realization of interfaces with DIH equipment. Provision of test benches (for actuation, perception, etc.).

- Technical support on scale-up:

Description of activity, value proposition (what?): Demonstration in real environment or in realistic conditions. Adaptation of existing technologies to specific use cases.

- Testing and validation:

Description of activity, value proposition (what?): Research aiming to reach TRL 5, removal of doubts, research and development, validation in lab.

- General information on technology services:

Format (how?): Collaborative research with several partners in Greece and EU for 1 to 5 years depending on the project. Conducting experiments with the assistance of DIH personnel.

Profile of DIH contact (who?): Project managers, senior research engineers, junior research engineers.

Involvement, effort from DIH: Support from experts in the submission of European or national proposals. Long term participation (typically more than 2 years but there are cases of annual projects) of several experts from the DIH for technical research and development in robotics.

Funding, business plan: Subsidized by the DIH (as part of investigating project opportunities).

- Financial charge for customer: Free or part of project investigation.

SINTEF / RINVE

RINVE is the Norwegian national network on I&M robotics, founded by SINTEF. It has about 30 members.

BREAKDOWN OF SERVICES

1. Ecosystem services

Awareness creation	
Category	Ecosystem service, Community building.
Description of activity, value proposition (what?)	Robotics awareness presentation for decision-makers and suppliers.
Format (how?)	Webinar/workshops/conferences where the DIH is invited to speak.
Profile of DIH contact (who?)	Senior technical experts.
Involvement, effort from DIH	Presentation by one technical expert.
Financial charge for customer	Free.
Funding, business plan	Subsidized by the DIH.
Dissemination	
Category	Ecosystem service, Community building.
Description of activity, value proposition (what?)	Communication of the services offered by the DIH. Dissemination of results and key facts in local language and in English.
Format (how?)	Information distributed through e.g. web sites, social media, newsletters, email.
Profile of DIH contact (who?)	Experts in communication, technical experts.
Involvement, effort from DIH	Involvement of the communication teams belonging to the organizations involved in the DIH.
Financial charge for customer	Free.
Funding, business plan	Subsidized by the DIH.

Ecosystem building	
Category	Ecosystem service, Community building.
Description of activity, value proposition (what?)	Putting people in touch with professional and sectorial associations, certification bodies, public and private investors in the region and in Europe.
Format (how?)	Reference information provided through the DIH web portal and via DIH contact points.
Profile of DIH contact (who?)	Business facilitators, technical experts.
Involvement, effort from DIH	Access to networks, clusters, associations and federations to which the organizations of the DIH belong and/or have contacts with.
Financial charge for customer	Free.
Funding, business plan	Subsidized by the DIH. Potential economic upside for DIH in generating projects which involve the DIH.

Brokerage, Matchmaking	
Category	Ecosystem service, Community building, Ecosystem learning.
Description of activity, value proposition (what?)	Organization of meetings and workshops between players in the value chain to bring demand and offer sides together to share ideas and analyse opportunities for collaboration.
Format (how?)	One event per year at a central location (e.g., an airport) or online with presentations from stakeholders, networking lunch (if not digital), dialogue B2B, tour of the DIH if visit on site.
Profile of DIH contact (who?)	Business facilitators, technical experts.
Involvement, effort from DIH	One day event with participation of 3 to 4 experts from the DIH, supply of communication material.
Financial charge for customer	Free.
Funding, business plan	Subsidized by the DIH, co-financed by the Research Council of Norway.
Mentoring by technical and application experts	
Category	Ecosystem service, Strategy development
Description of activity, value proposition (what?)	Advise on potential uses of robotics and automation.
Format (how?)	Conference calls, bilateral meetings, meetings on customer's premises.
Profile of DIH contact (who?)	Technical and application experts.
Involvement, effort from DIH	One hour with experts.
Financial charge for customer	Free (typically 1-2 hours, online) or paid for larger efforts.
Funding, business plan	Subsidized by the DIH (as part of investigating project opportunities) or cost per hour for larger efforts. In some cases, a mix of public and private funding is used.
Resources and knowledge sharing	
Category	Ecosystem service, Ecosystem learning.
Description of activity, value proposition (what?)	Informing about and organizing access to resources, databases, seminars, workshops, conferences, field trips, newsletters and other information of interest.
Format (how?)	Reference information provided through the DIH, newsletters and RIMA network web portal. Conferences, field trips, workshops organized by the DIH or in cooperation with partners.
Profile of DIH contact (who?)	Communication team. Experts.
Involvement, effort from DIH	Involvement of communication teams.
Financial charge for customer	Free (information sharing, some conferences) or paid (e.g., conference fee)
Funding, business plan	Mixed public and private funding.

DIH network promotion, Promotion of the offer	
Category	Ecosystem service, Ecosystem learning.
Description of activity, value proposition (what?)	Promote both the DIH and its members.
Format (how?)	Webpages, newsletters. Participation on roadshows, exhibitions, dissemination, and other activities according to customer interests (e.g., suppliers, end users).
Profile of DIH contact (who?)	Communication, business and technical experts.
Involvement, effort from DIH	Involvement of the communication teams and technical experts.
Financial charge for customer	Free or paid.
Funding, business plan	Subsidized by the DIH. Mix of public and private funding.

2. Business services

Business development[4]	
Category	Incubator/accelerator support.
Description of activity, value proposition (what?)	Dedicated mostly to spin-off and start-up companies promoting SINTEF technologies. Support for business creation, Facilitating the creation of start-ups, protecting research results, advising about IP management, carrying out strategic analyses, finding scientific and technical information and business intelligence, setting up industrial and collaborative contracts.
Format (how?)	Meetings, conference calls, project management.
Profile of DIH contact (who?)	Experts in business creation, in IP management.
Involvement, effort from DIH	Meetings with experts in business development.
Financial charge for customer	Free for SINTEF spin-offs, otherwise paid.
Funding, business plan	Mixed public funding and internal DIH investment.
Connect to funding sources	
Category	Access to financing, Incubator/accelerator support.
Description of activity, value proposition (what?)	Connecting businesses and public organizations to relevant funding opportunities including tax refund mechanisms. Leading or contributing to proposal writing in response to funding opportunities. Business development as part of proposal writing.
Format (how?)	Meetings, conference calls, proposal writing.
Profile of DIH contact (who?)	Business developer, technical expert.

4 Regarding SINTE technologies

Involvement, effort from DIH	Meetings with experts in funding of RDI in robotics. Proposal writing.
Financial charge for customer	Free (sales activity, or as limited support offer) or paid (if the DIH involvement is only in proposal/funding investigation phase).
Funding, business plan	Subsidized by the DIH when part of sales activity or limited support offer (e.g., a 1-hour information meeting). Private funding when the DIH is involved only in the sales phase and not in the potential subsequent project.
Hands-on training	
Category	Skills and education.
Description of activity, value proposition (what?)	Training of technical and scientific personnel.
Format (how?)	Courses given by DIH personnel. On demand and regular.
Profile of DIH contact (who?)	Technical experts.
Involvement, effort from DIH	1-2 day events and shorter events, supply of training reference material.
Financial charge for customer	Yes.
Funding, business plan	Flat fee per seminar/training course.
Research, development and innovation opportunities	
Category	Offering housing.
Description of activity, value proposition (what?)	Facilitating, hosting, and mentoring early-phase companies spun off from SINTEF.
Format (how?)	On demand.
Profile of DIH contact (who?)	Business facilitator.
Involvement, effort from DIH	Office space and bench space located in DIH premises, Internet infrastructure, lab facilities.
Financial charge for customer	Free.
Funding, business plan	Subsidized by SINTEF (e.g., the research group from which the personnel entering the spin-off originates from).
Hosting spin-off and start-up companies	
Category	Skills and education.
Description of activity, value proposition (what?)	Training of technical and scientific personnel.
Format (how?)	Courses given by DIH personnel. On demand and regular.
Profile of DIH contact (who?)	Technical experts.
Involvement, effort from DIH	1-2 day events and shorter events, supply of training reference material.
Financial charge for customer	Yes.
Funding, business plan	Flat fee per seminar/training course.

3. Technology services

Research and development collaboration	
Category	Strategic RDI.
Description of activity, value proposition (what?)	Research aiming to reach TRL 5, removal of doubts, specifications, research and development, validation in laboratory.
Format (how?)	Collaborative research with several partners in Norway, EU and/or outside the EU for 2 years or more.
Profile of DIH contact (who?)	Project manager, technical experts.
Involvement, effort from DIH	Support from experts in the submission of European or national proposals. Long-term participation (over 2 years) of several experts from the DIH for technical research and development in robotics.
Financial charge for customer	Free or paid and/or via in-kind contributions from the compnay(s).
Funding, business plan	Funded by European or national research programmes, mix of public and private funding, or solely private funding (in rare cases).
Innovation projects: technology demonstrators, proof of concepts	
Category	Contract research.
Description of activity, value proposition (what?)	Research and development aiming to reach TRL 6, application of existing technologies, validation in laboratory.
Format (how?)	Bilateral or collaborative projects from one to a few years.
Profile of DIH contact (who?)	Project manager, technical experts.
Involvement, effort from DIH	Support from experts in the submission of European or national proposals. Participation of several experts from the DIH for technical research and development in robotics.
Financial charge for customer	Yes.
Funding, business plan	Funded by bilateral contracts, European or national innovation programmes, private investors.
Proof of concept, prototyping, commercialization/industrialization, small series production	
Category	Technical support on scale-up.
Description of activity, value proposition (what?)	Demonstration in real environment or in realistic conditions. Adaptation of existing technologies to the specific use case.
Format (how?)	Bilateral projects from one to several months
Profile of DIH contact (who?)	Technical experts.
Involvement, effort from DIH	Support from experts in the submission of European or national proposals. Participation of several experts from the DIH for development and innovation.
Financial charge for customer	Yes.
Funding, business plan	Funded by bilateral contracts, European or national innovation programmes, investors.

Lab facilities, equipment for lease/use (sensors, robots, robot tooling, etc.), rapid prototyping	
Category	Provision of technological infrastructure.
Description of activity, value proposition (what?)	Demonstration with equipment held in DIH laboratories or deployment of sensors etc. at customer site. Realization of interfaces with DIH equipment. Provision of test benches (for actuation, perception, etc.). Provision of large-scale field laboratories (e.g., test area for autonomous ships in the Trondheim fjord, or an aquaculture facility).
Format (how?)	Conducting experiments with the assistance of DIH personnel.
Profile of DIH contact (who?)	Technical experts.
Involvement, effort from DIH	Support from experts in the submission of European or national proposals. Participation of several experts from the DIH for maturation.
Financial charge for customer	Yes: financial charge based on daily rate and depending on the personnel involved, equipment used at the DIH and the amount of adaptation needed.
Funding, business plan	Funded by bilateral contracts, European or national innovation programmes, investors.
Certification, product demonstration & testing, comparison of products	
Category	Testing and validation.
Description of activity, value proposition (what?)	For testing and benchmarking, SINTEF holds and can offer access to a set of dedicated to infrastructures and test beds to perform assessment under realistic conditions involving e.g., aquaculture, pipelines, high-voltage environment, outdoor/indoor, large-scale pools, ocean (outdoor) facilities). SINTEF can also organize access to infrastructure in other application sectors.
Format (how?)	Conducting experiments with the assistance of DIH personnel following evaluation protocols. Provision of metrics and tools for benchmarking the utility, usability, effectiveness.
Profile of DIH contact (who?)	Technical experts. For certification issues, SINTEF usually involves external parties.
Involvement, effort from DIH	Support from experts in the submission of European or national proposals. Participation of several experts from the DIH for maturation.
Financial charge for customer	Yes.
Funding, business plan	Funded by bilateral contracts, European or national innovation programmes, investors.

The Technical University of Košice (TUKE)

BREAKDOWN OF SERVICES

1. Ecosystem services

Awareness creation	
Category	Ecosystem service: Community building.
Description of activity, value proposition (what?)	Robotics awareness presentations for decision-makers and suppliers.
Format (how?)	Webinar/workshops/conferences with the DIH as an invited speaker.
Profile of DIH contact (who?)	Senior technical experts.
Involvement, effort from DIH	Presentation by one technical expert.
Financial charge for customer	Free.
Funding, business plan	Subsidized by the DIH.

Dissemination	
Category	Ecosystem service: Community building.
Description of activity, value proposition (what?)	Communication of the services offered by the DIH. Dissemination of results and key facts in the local language and in English.
Format (how?)	Information distributed through e.g. web sites, social media, newsletters, email.
Profile of DIH contact (who?)	Experts in communication, technical experts.
Involvement, effort from DIH	Involvement of the communication teams belonging to the organizations involved in the DIH.
Financial charge for customer	Free.
Funding, business plan	Subsidized by the DIH.
Ecosystem building	
Category	Ecosystem service: Community building.
Description of activity, value proposition (what?)	Putting people in touch with professional and sectorial associations, certification bodies, public and private investors in region and in Europe.
Format (how?)	Reference information provided through the DIH web portal and via DIH contact points.
Profile of DIH contact (who?)	Business facilitators, technical experts.
Involvement, effort from DIH	Access to networks, clusters, associations and federations to which the organizations of the DIH belong and/or have contacts with.
Financial charge for customer	Free.
Funding, business plan	Subsidized by the DIH. Potential economic upside for DIH in generating projects which involve the DIH.

Brokerage, Matchmaking	
Category	Ecosystem service: Community building
Description of activity, value proposition (what?)	Organization of meetings and workshops between players in the value chain to bring demand and offer sides together to share ideas and analyse opportunities for collaboration.
Format (how?)	Minimum one event per year at the DIH or online with presentations from stakeholders, networking lunch (if not digital), dialogue B2B, tour of the DIH if the visit takes place on site.
Profile of DIH contact (who?)	Business facilitators, technical experts.
Involvement, effort from DIH	One day event with participation of 3 to 4 experts from the DIH, supply of communication material.
Financial charge for customer	Free.
Funding, business plan	Subsidized by the DIH.
Market research, market awareness, market/technology trends	
Category	Ecosystem service: Strategy development.
Description of activity, value proposition (what?)	Provision of access to experts in market analysis, market data held in the network and at the DIH at European and international level.
Format (how?)	Conference call, meeting with business developers at the DIH.
Profile of DIH contact (who?)	Contact to experts in market analysis.
Involvement, effort from DIH	One hour with experts.
Financial charge for customer	Yes.
Funding, business plan	Subsidized by the DIH.

Mentoring by technical and application experts	
Category	Ecosystem service: Strategy development.
Description of activity, value proposition (what?)	Advise on potential uses of robotics and automation.
Format (how?)	Conference calls, bilateral meetings, meetings on customer's premises.
Profile of DIH contact (who?)	Technical and application experts.
Involvement, effort from DIH	One hour with experts.
Financial charge for customer	Free (typically 1-2 hours, online) or by payment for larger efforts.
Funding, business plan	Subsidized by the DIH (as part of investigating project opportunities) or cost per hour for larger efforts. In some cases, a mix of public and private funding is used.
Resources and knowledge sharing	
Category	Ecosystem service: Ecosystem learning.
Description of activity, value proposition (what?)	Informing and organizing access to resources, databases, seminars, workshops, conferences, field trips, newsletters and other information of interest.

Format (how?)	Reference information provided through the DIH, newsletters and RIMA network web portal. Conferences, field trips, workshops organized by the DIH or in cooperation with partners.
Profile of DIH contact (who?)	Communication team. Experts.
Involvement, effort from DIH	Involvement of the communication teams of the organizations involved in the DIH.
Financial charge for customer	Free (information sharing, some conferences) or paid (e.g., conference fee)
Funding, business plan	Mixed public and private funding.

DIH network promotion, promotion of the offer	
Category	Ecosystem service: Ecosystem learning.
Description of activity, value proposition (what?)	Promote both the DIH and its members.
Format (how?)	Webpages, newsletters. Participation in roadshows, exhibitions, dissemination, and other activities according to customer interests (e.g., suppliers, end users).
Profile of DIH contact (who?)	Communication, business and technical experts.
Involvement, effort from DIH	Involvement of the communication teams and technical experts.
Financial charge for customer	Free or paid.
Funding, business plan	Subsidized by the DIH. Mix of public and private funding.

2. Business services

Business development[5]	
Category	Business service: Incubator/accelerator support.
Description of activity, value proposition (what?)	Dedicated mostly to spin-off and start-up companies promoting TUKE technologies. Support for business creation, facilitating the creation of start-ups, protecting research results, advising on IP management, carrying out strategic analyses, finding scientific and technical information, business intelligence, setting up industrial and collaborative contracts.
Format (how?)	Meetings, conference calls, project management.
Profile of DIH contact (who?)	Experts in business creation, in IPR management.
Involvement, effort from DIH	Meetings with experts in business development.
Financial charge for customer	Free for TUKE member.
Funding, business plan	Mixed public funding and internal investment in DIH.

5 Regarding TUKE technologies

Connect to funding sources	
Category	Business service: Access to financing.
Description of activity, value proposition (what?)	Connect businesses and public organizations to relevant funding opportunities including tax refund mechanisms. Lead or contribute to proposal writing in response to funding opportunities. Business development as part of proposal writing.
Format (how?)	Meetings, conference calls, proposal writing.
Profile of DIH contact (who?)	Business developer, technical expert.
Involvement, effort from DIH	Meetings with experts in funding of research, development and innovation in robotics. Proposal writing.
Financial charge for customer	Free (sales activity, or as limited support offer) or paid (DIH involvement only in proposal/funding investigating phase)
Funding, business plan	Subsidized by the DIH when as part of sales activity or as limited support offer (e.g., a 1-hour information meeting). Private funding when the DIH is involved only in the sales phase and not in the potential subsequent project.
Hands-on training	
Category	Business service: Skills and education.
Description of activity, value proposition (what?)	Training of technical and scientific personnel.
Format (how?)	Courses given by DIH personnel. On demand and regular.
Profile of DIH contact (who?)	Technical experts.
Involvement, effort from DIH	1-2 day events and shorter events, supply of training reference material.
Financial charge for customer	Yes.
Funding, business plan	Flat fee per seminar/training course.
Research, development and innovation opportunities	
Category	Business service: Project development.
Description of activity, value proposition (what?)	Advising on technological research, development and innovation activity. Elaboration of an action plan for RDI with milestones, deliverables, agreement on the respective role of stakeholders and on the final delivery. Basis for initiating, e.g., bilateral projects between industry and RTOs in the DIH, or as basis for proposal writing to secure funding.
Format (how?)	Meetings, conference calls.
Profile of DIH contact (who?)	Business developer, technical expert.
Involvement, effort from DIH	Meetings with experts in research, development and innovation in robotics.
Financial charge for customer	Free if the activity can lead to project for the DIH or else paid.

Funding, business plan	Subsidized by the DIH. Private finding.
Hosting spin-off and start-up companies	
Category	Business service: Offering housing.
Description of activity, value proposition (what?)	Facilitating, hosting, and mentoring early-phase companies spun off from TUKE.
Format (how?)	On demand.
Profile of DIH contact (who?)	Business facilitator.
Involvement, effort from DIH	Office space, workspace, located in DIH premises, Internet infrastructure, lab facilities.
Financial charge for customer	Free.
Funding, business plan	Paid or granted by national or EU Funds.

3. Technology services

Research and development collaboration	
Category	Technology service: Strategic RDI.
Description of activity, value proposition (what?)	Research aiming to reach TRL 5/6, removal of doubts, specifications, research and development, validation in laboratory.
Format (how?)	Collaborative research with several partners in Slovakia, EU and/or outside the EU for 2 years or more.
Profile of DIH contact (who?)	Project manager, technical experts.
Involvement, effort from DIH	Support from experts in the submission of European or national proposals. Long-term participation (over 2 years) of several experts from the DIH for technical research and development in robotics.
Financial charge for customer	Free or paid (possibly in kind) by industry.
Funding, business plan	Funded by European or national research programmes, mix of public and private funding, or solely private funding (rare cases).
Innovation projects: technology demonstrators, proof of concepts	
Category	Technology service: Contract research.
Description of activity, value proposition (what?)	Research and development aiming to reach TRL 6, application of existing technologies, validation in laboratory.
Format (how?)	Bilateral or collaborative projects from one to a few years.
Profile of DIH contact (who?)	Project manager, technical experts.
Involvement, effort from DIH	Support from experts in the submission of European or national proposals. Participation of several experts from the DIH for technical research and development in robotics
Financial charge for customer	Yes.
Funding, business plan	Funded by bilateral contracts, European or national innovation programmes, private investors.

Proof of concept, prototyping, commercialization/industrialization, small series production	
Category	Technical service: technical support.
Description of activity, value proposition (what?)	Demonstration in real environment or in realistic conditions. Adaptation of existing technologies to specific use case.
Format (how?)	Bilateral projects from one to several months.
Profile of DIH contact (who?)	Technical experts.
Involvement, effort from DIH	Support from experts in the submission of European or national proposals. Participation of several experts from the DIH for development and innovation
Financial charge for customer	Yes.
Funding, business plan	Funded by bilateral contracts, European or national innovation programmes, investors.
Lab facilities, equipment for lease/use (sensors, robots, robot tooling, etc.), rapid prototyping	
Category	Technical service: Provision of technological infrastructure.
Description of activity, value proposition (what?)	Demonstration with equipment held in the Prototyping & Innovation Centre or laboratories at the DIH or deployment of sensors etc. at customer site. Realization of interfaces with DIH equipment. Provision of test benches (for actuation, perception, etc.).
Format (how?)	Conducting experiments with the assistance of DIH personnel.
Profile of DIH contact (who?)	Technical experts.
Involvement, effort from DIH	Support from experts in the submission of European or national proposals. Participation of several experts from the DIH for maturation.
Financial charge for customer	Yes: financial charge based on daily rate and depending on the personnel involved, equipment used at the DIH and adaptation needed.
Funding, business plan	Funded by bilateral contracts, European or national innovation programmes, investors.
Certification, product demonstration & testing, comparison of products	
Category	Technical service: Testing and validation.
Description of activity, value proposition (what?)	For testing and benchmarking, TUKE can offer access to a set of dedicated infrastructures (e.g., aquaculture, pipelines, high-voltage environment, outdoor/indoor, large-scale pools, ocean (outdoor) facilities), where technologies can be assessed under realistic conditions. TUKE can also organize access to infrastructure in other application sectors.

Format (how?)	Conducting experiments with the assistance of DIH personnel following evaluation protocols. Provision of metrics and tools for benchmarking the utility, usability, effectiveness.
Profile of DIH contact (who?)	Technical experts. For certification issues, TUKE usually involves the Technical Inspection Office in Slovakia.
Involvement, effort from DIH	Support from experts in the submission of European or national proposals. Participation of several experts from the DIH for maturation.
Financial charge for customer	Yes.
Funding, business plan	Funded by bilateral contracts, European or national innovation programmes, investors.

i-Botics

The i-Botics DIH comprises TNO and the University of Twente.

BREAKDOWN OF SERVICES

1. Ecosystem services

Awareness creation	
Category	Ecosystem service, Community building.
Description of activity, value proposition (what?)	Robotics awareness presentation for decision-makers and suppliers.
Format (how?)	Webinar/workshops/conferences that the DIH is invited to, on national or European level.
Profile of DIH contact (who?)	Senior technical experts.
Involvement, effort from DIH	Presentation by one technical expert.
Financial charge for customer	Free.
Funding, business plan	Subsidized by the DIH.

Dissemination	
Category	Ecosystem service, Community building.
Description of activity, value proposition (what?)	Communication of the services offered by the DIH. Dissemination of results and key facts in local language and in English.
Format (how?)	Information distributed through e.g. web sites, social media, newsletters, email.
Profile of DIH contact (who?)	Experts in communication, technical experts.
Involvement, effort from DIH	Involvement of the communication teams belonging to the organizations involved in the DIH.
Financial charge for customer	Free.
Funding, business plan	Subsidized by the DIH.
Ecosystem building	
Category	Ecosystem service, Community building.
Description of activity, value proposition (what?)	Putting people in touch with professional and sectorial associations, certification bodies, public and private investors in region and in Europe.
Format (how?)	Reference information provided through the DIH web portal and via DIH contact points.
Profile of DIH contact (who?)	Business facilitators, technical experts.
Involvement, effort from DIH	Provision of access to networks, clusters, associations and federations to which the organizations of the DIH belong and/or have contact with.
Financial charge for customer	Free.
Funding, business plan	Subsidized by the DIH. Potential economic upside for DIH in generating projects which involve the DIH.

Brokerage, Matchmaking	
Category	Ecosystem service, Community building, Ecosystem learning.
Description of activity, value proposition (what?)	Organization of meetings and workshops between players in the value chain to bring demand and offer sides together to share ideas and analyse opportunities for collaboration.
Format (how?)	On demand, on average 1 or 2 times a year, either at TNO, at the UT or online with presentations from stakeholders, networking lunch or dinner (if not digital), dialogue B2B, tour of the DIH if visit on site.
Profile of DIH contact (who?)	Business facilitators, technical experts, students.
Involvement, effort from DIH	Half-day event with participation of 3 to 4 experts from the DIH, supply of communication material.
Financial charge for customer	Free.
Funding, business plan	Subsidized by the DIH.
Market research, market awareness, market/technology trends	
Category	Ecosystem service, Strategy development, Consulting.
Description of activity, value proposition (what?)	Provision of access to experts in market analysis, market data held in the network and at the DIH in European and International level.
Format (how?)	Conference call, meeting with business developers at the DIH.
Profile of DIH contact (who?)	Contact to experts in market analysis.
Involvement, effort from DIH	One hour with experts.
Financial charge for customer	Sometimes paid, sometimes free.
Funding, business plan	Subsidized by the DIH.

Mentoring by technical and application experts	
Category	Ecosystem service, Strategy development.
Description of activity, value proposition (what?)	Advise on potential uses of robotics and automation.
Format (how?)	Conference calls, bilateral meetings, meetings on customer's premises.
Profile of DIH contact (who?)	Technical and application experts.
Involvement, effort from DIH	One hour with experts.
Financial charge for customer	Free (typically 1-2 hours, online) or by payment for larger efforts.
Funding, business plan	Subsidized by the DIH (as part of investigating project opportunities) or cost per hour for larger efforts. In some cases, a mix of public and private funding is used.

Resources and knowledge sharing	
Category	Ecosystem service, Ecosystem learning.
Description of activity, value proposition (what?)	Informing about and organizing access to resources, databases, seminars, workshops, conferences, field trips, newsletters and other information of interest
Format (how?)	Reference information provided through the DIH, newsletters and RIMA network web portal. Conferences, field trips, workshops organized by the DIH or in cooperation with partners.
Profile of DIH contact (who?)	Communication team. Experts.
Involvement, effort from DIH	Involvement of the communication teams of the organizations involved in the i-Botics DIH.
Financial charge for customer	Free (information sharing, some conferences) or paid (e.g., conference fee)
Funding, business plan	Mixed public and private funding.

DIH network promotion, Promotion of the offer	
Category	Ecosystem service, Ecosystem learning.
Description of activity, value proposition (what?)	Promote both the DIH and its members.
Format (how?)	Webpages, newsletters. Participation in roadshows, exhibitions, dissemination, and other activities according to customer interests (e.g., suppliers, end users).
Profile of DIH contact (who?)	Communication, business and technical experts.
Involvement, effort from DIH	Involvement of the communication teams and technical experts.
Financial charge for customer	Free or paid.
Funding, business plan	Subsidized by the DIH. Mix of public and private funding.

2. Business services

Business development[6]	
Category	Incubator/accelerator support.
Description of activity, value proposition (what?)	Dedicated mostly to spin-off and start-up companies promoting TNO/UT technologies. Support for business creation, facilitating the creation of start-ups, protecting research results, advising about IP management, carrying out strategic analyses, finding scientific and technical information, business intelligence, setting up industrial and collaborative contracts.
Format (how?)	Meetings, conference calls, project management.

[6] Regarding TNO/UT technologies

Profile of DIH contact (who?)	Experts in business creation and in IPR management.
Involvement, effort from DIH	Presentation by one technical expert.
Financial charge for customer	Reviewed per case.
Funding, business plan	Reviewed per case.
Connect to funding sources	
Category	Access to financing, Incubator/accelerator support.
Description of activity, value proposition (what?)	Connect businesses and public organizations to relevant funding opportunities. Lead or contribute to proposal writing in response to funding opportunities. Business development as part of proposal writing. This can also involve forming consortia from the existing community.
Format (how?)	Meetings, conference calls, proposal writing.
Profile of DIH contact (who?)	Business developer, technical expert.
Involvement, effort from DIH	Meetings with experts in funding of research, development and innovation in robotics. Proposal writing.
Financial charge for customer	Free (if DIH is included in the consortium) or paid (DIH involvement only in proposal/funding investigating phase)
Funding, business plan	Subsidized by the DIH when part of a sales activity or a limited support offer (e.g., a 1-hour information meeting). Reviewed for eligibility when DIH involvement is only in sales phase and not in the potentially resulting project.
Hands-on training	
Category	Skills and education.
Description of activity, value proposition (what?)	Training of technical and scientific personnel.
Format (how?)	Courses given by DIH personnel. On demand and regular.
Profile of DIH contact (who?)	Technical experts.
Involvement, effort from DIH	1-2 day events and shorter events, supply of training reference material.
Financial charge for customer	Yes.
Funding, business plan	Flat fee per seminar/training course.
Research, development and innovation opportunities	
Category	Project development, Access to finance.
Description of activity, value proposition (what?)	Advice on technological research, development and innovation activity. Elaboration of an action plan for RDI with milestones, deliverables, agreement on the respective role of stakeholders and on the final delivery. Basis for initiating, e.g., bilateral projects between industry and RTOs in the DIH, or as a basis for proposal writing to secure funding.

Format (how?)	Meetings, conference calls.
Profile of DIH contact (who?)	Business developer, technical expert.
Involvement, effort from DIH	Meetings with experts in research, development and innovation in robotics.
Financial charge for customer	Free if the activity can lead to a project for the DIH, otherwise paid.
Funding, business plan	Subsidized by the DIH. Private finding.
Hosting spin-off and start-up companies	
Category	Offering housing.
Description of activity, value proposition (what?)	Facilitating, hosting, and mentoring early-phase companies spun off from UT.
Format (how?)	On demand.
Profile of DIH contact (who?)	Business facilitator.
Involvement, effort from DIH	Office space, workspace, located in DIH premises, internet infrastructure, lab facilities.
Financial charge for customer	Depends on the case.
Funding, business plan	Depends on the case.

3. Technology services

Research and development collaboration	
Category	Strategic RDI.
Description of activity, value proposition (what?)	Research aiming to reach TRL 5, removal of doubts, specifications, research and development, validation in laboratory.
Format (how?)	Collaborative research with several partners in the Netherlands, EU and/or outside the EU for 2 years or more.
Profile of DIH contact (who?)	Project manager, technical experts.
Involvement, effort from DIH	Support from experts in the submission of European or national proposals. Long-term participation (over 2 years) of several experts from the DIH for technical research and development in robotics.
Financial charge for customer	Free or paid (sometimes in kind).
Funding, business plan	Funded by European or national research programmes, mix of public and private funding, or solely private funding (rare cases).
Innovation projects: technology demonstrators, proof of concepts	
Category	Contract research.
Description of activity, value proposition (what?)	Research and development aiming to reach TRL 6, application of existing technologies, validation in laboratory.
Format (how?)	Bilateral or collaborative projects from one to a few years.
Profile of DIH contact (who?)	Project manager, technical experts.
Involvement, effort from DIH	Support from experts in the submission of European or national proposals. Participation of several experts from the DIH for technical research and development in robotics

Financial charge for customer	Yes.
Funding, business plan	Funded by bilateral contracts, European or national innovation programmes, private investors.
Proof of concept, prototyping, commercialisation/industrialisation, small series production	
Category	Technical support on scale-up.
Description of activity, value proposition (what?)	Demonstration in real environment or in realistic conditions. Adaptation of existing technologies to the specific use case.
Format (how?)	Bilateral projects from one to several months.
Profile of DIH contact (who?)	Technical experts.
Involvement, effort from DIH	Support from experts in the submission of European or national proposals. Participation of several experts from the DIH for development and innovation.
Financial charge for customer	Yes.
Funding, business plan	Funded by bilateral contracts, European or national innovation programmes, investors.
Lab facilities, equipment for lease/use (sensors, robots, robot tooling, etc.), rapid prototyping	
Category	Provision of technological infrastructure.
Description of activity, value proposition (what?)	Demonstration with equipment held in the laboratories at the DIH or deployment of sensors etc. at customer site. Realization of interfaces with DIH equipment. Provision of test benches (for actuation, perception, etc.). Visit websites of the DIH partners or contact the DIH to check the possibilities.
Format (how?)	Conducting experiments with the assistance of DIH personnel.
Profile of DIH contact (who?)	Technical experts.
Involvement, effort from DIH	Support from experts in the submission of European or national proposals. Participation of several experts from the DIH for maturation.
Financial charge for customer	Yes, financial charge based on daily rate and the personnel involved, equipment used at the DIH and adaptation needed.
Funding, business plan	Funded by bilateral contracts, European or national innovation programmes, investors.

VTT and SMACC

The Technical Research Centre of Finland (VTT) and SMACC (Smart machines and manufacturing competence centre) DIH gathers several legal entities: VTT, Tampere University (TUNI), Tampere University of Applied Sciences (TAMK), Tamlink ltd.

BREAKDOWN OF SERVICES

1. Ecosystem services

Awareness creation	
Category	Ecosystem service, Community building.
Description of activity, value proposition (what?)	Robotics awareness, advances and possibilities info for decision-makers.
Format (how?)	Project sales, proposals, workshops, webinars (sales activity)
Profile of DIH contact (who?)	Sales personnel, Technical experts.
Involvement, effort from DIH	One half-day event with senior experts from DIH explaining the project.
Financial charge for customer	Free.
Funding, business plan	Subsidized by the DIH, aiming to sell the project.

Dissemination	
Category	Ecosystem service, Community building.
Description of activity, value proposition (what?)	Communication about the offer of services from the DIH. Dissemination of the references, results and key facts in local language and in English.
Format (how?)	Information distributed through e.g. web sites, social media, newsletters, email.
Profile of DIH contact (who?)	Experts in communication.
Involvement, effort from DIH	Involvement of the communication team of VTT and other DIH organizations.
Financial charge for customer	Free.
Funding, business plan	Subsidized by the DIH.
Ecosystem building	
Category	Ecosystem service, Community building.
Description of activity, value proposition (what?)	Allowing people to get in touch with sectorial associations, certification bodies and public and private investors in the region and in Europe.
Format (how?)	Reference information provided through the DIH web portal. Access to experts.
Profile of DIH contact (who?)	Business facilitators, technical experts.
Involvement, effort from DIH	Involvement of the communication team of VTT and other DIH organizations.
Financial charge for customer	Free.
Funding, business plan	Subsidized by the DIH.

Brokerage, Matchmaking	
Category	Ecosystem service, Community building.
Description of activity, value proposition (what?)	Organization of meetings between players in the value chain. Discussing interfaces and information exchange between players. Share ideas and analyse opportunities
Format (how?)	One meeting within each experiment (or more if necessary). With presentations from stakeholders, visit to the DIH, possibly networking lunch.
Profile of DIH contact (who?)	Business facilitators, technical experts.
Involvement, effort from DIH	One-day event with participation of 3 to 4 experts from the DIH, supply of communication material.
Financial charge for customer	Free (as a part of the project)
Funding, business plan	Subsidized by the DIH.
Market research, market awareness, market/technology trends	
Category	Ecosystem service, Strategy development.
Description of activity, value proposition (what?)	Market analysis and market data by the experts from the network and the DIH at European and international level.
Format (how?)	Meeting with business developers at the DIH.
Profile of DIH contact (who?)	DIH experts in market analysis.
Involvement, effort from DIH	Required time with experts.
Financial charge for customer	Yes (if not part of the experiment)
Funding, business plan	Subsidized by the DIH.
Technical consultancy	
Category	Ecosystem service, Strategy development.
Description of activity, value proposition (what?)	Consultation/advising on technical questions on robotics and automation (or other technical expert problems)
Format (how?)	Conference calls, meetings, projects.
Profile of DIH contact (who?)	Technical and application experts.
Involvement, effort from DIH	Discussions or projects with experts.
Financial charge for customer	Yes (if not part of the experiment)
Funding, business plan	Subsidized by the DIH.

Resources and knowledge sharing	
Category	Ecosystem service, Ecosystem learning.
Description of activity, value proposition (what?)	Informing about and enabling access to resources, seminars, workshops, conferences, newsletters, and other information of interest.
Format (how?)	Information provided through the DIH and RIMA network web portal. Possible access to experts for advanced consultancy.
Profile of DIH contact (who?)	DIH contact person for communication and/or innovation.
Involvement, effort from DIH	The communication team of the DIH organizations.

Financial charge for customer	Free, if part of the project or marketing actions.
Funding, business plan	Mixed public and private funding.
DIH network promotion, Promotion of the offer	
Category	Ecosystem service, Ecosystem learning.
Description of activity, value proposition (what?)	Promote the DIH, its members and services.
Format (how?)	Organization of campaigns on social media, newspapers and professional magazines. Participation in fairs and events.
Profile of DIH contact (who?)	Communication, business and technical experts.
Involvement, effort from DIH	Communication teams of the DIH.
Financial charge for customer	Free.
Funding, business plan	Subsidized by the DIH.

2. Business services

Business development	
Category	Incubator/accelerator support.
Description of activity, value proposition (what?)	Support for the business creation of spin-off and start-up companies, facilitating the creation of start-ups, protecting research results, advising about IP management, carrying out strategic analyses, finding scientific and technical information, business intelligence, setting up industrial and collaborative contracts.
Format (how?)	However is agreed.
Profile of DIH contact (who?)	Experts in business creation, in IP management.
Involvement, effort from DIH	Meetings with experts in business development.
Financial charge for customer	Free for DIH members.
Funding, business plan	Subsidized by the DIH.
Funding opportunities investigation	
Category	Incubator/accelerator support.
Description of activity, value proposition (what?)	Advice about: opportunities offered by European research and innovation programmes, regional sources of funding, tax refund mechanisms, and private investors. Dedicated to spin-off and start-up companies promoting VTT or other DIH member technologies. Support for business creation, facilitating the creation of start-ups, protecting research results, advising about IP management, carrying out strategic analyses, finding scientific and technical information, obtaining business intelligence, setting up industrial and collaborative contracts.

Format (how?)	As agreed, meetings, conference calls.
Profile of DIH contact (who?)	Business developer, technical expert, experts in IP management.
Involvement, effort from DIH	Meetings with experts in business development.
Financial charge for customer	Meetings with experts in funding of RDI in robotics.
Funding, business plan	Subsidized by the DIH.
Hands-on training	
Category	Skills and education.
Description of activity, value proposition (what?)	Training operators in teleoperation and robotics.
Format (how?)	Courses and hands-on training on equipment available at the DIH. On demand.
Profile of DIH contact (who?)	Training by university training organization.
Involvement, effort from DIH	Not planned yet, future activities.
Financial charge for customer	Yes.
Funding, business plan	Rates according to the trainer's profile.
Research, development and innovation opportunities	
Category	Project development.
Description of activity, value proposition (what?)	Dedicated to organizations interested in promoting new technologies or willing to do collaborative research with VTT or other DIH members. Cooperative planning of technological research, finding partners needed for development and innovation activities. Elaboration of an action plan for RDI with milestones, deliverables, agreements on the respective role of stakeholders and on the final delivery.
Format (how?)	Planning the project in meetings, conference calls etc. Running the project in company and DIH laboratories as planned and needed.
Profile of DIH contact (who?)	Business developer, technical expert.
Involvement, effort from DIH	Meetings with experts in RDI in robotics.
Financial charge for customer	Free as part of the project.
Funding, business plan	Subsidized by the DIH.
Hosting spin-off and start-up companies	
Category	Offering housing.
Description of activity, value proposition (what?)	Accompanying, hosting, and mentoring spin-off or start-up companies willing to use and be located in DIH facilities.
Format (how?)	On demand.
Profile of DIH contact (who?)	Business facilitator.
Involvement, effort from DIH	Office space, workspace, located in DIH premises.
Financial charge for customer	Yes.
Funding, business plan	Rates according to needs.

3. Technology services

Research and development collaboration	
Category	Strategic RDI.
Description of activity, value proposition (what?)	PoC research reaching TRL 5, specifications, research and development, validation in laboratory.
Format (how?)	Research with Business Finland funding using their funding models, OR collaborative research with several partners in the EU for 2 years or more.
Profile of DIH contact (who?)	Technical experts, others if needed.
Involvement, effort from DIH	Support from DIH experts in the submission of European or national proposals. Long- term participation (over 2 years) of several experts from the DIH for technical research and development in robotics.
Financial charge for customer	Free (if funded by national funding)
Funding, business plan	Funded by European or national research programmes.
Innovation projects: technology demonstrators, proof of concepts	
Category	Contract research.
Description of activity, value proposition (what?)	Research and development aiming to develop applications for international markets. Validation in laboratory.
Format (how?)	Bilateral projects from one to a few years.
Profile of DIH contact (who?)	Project manager, technical experts.
Involvement, effort from DIH	Support from experts in the submission of European or national proposals. Long-term participation (over 2 years) of several experts from the DIH for technical research and development in robotics.
Financial charge for customer	Yes (national funding applied)
Funding, business plan	Funded by bilateral contracts, European or national innovation programmes, private investors.
Proof of concept, prototyping, commercialization/industrialization, small series production	
Category	Technical support on scale-up.
Description of activity, value proposition (what?)	Demonstration in real environment or in realistic conditions. Adaptation of existing technologies to the specific use case.
Format (how?)	Nationally-funded project lasting one to several months.
Profile of DIH contact (who?)	Technical experts.
Involvement, effort from DIH	Support from experts in the submission of European or national proposals. Participation of several experts from the DIH for development and innovation.
Financial charge for customer	Partly customer-funded, partly national or European funding

Funding, business plan	Funded by bilateral contracts, European or national innovation programmes, investors.
Lab facilities, equipment for lease/use (sensors, robots, robot tooling, etc.), rapid prototyping	
Category	Provision of technological infrastructure.
Description of activity, value proposition (what?)	Demonstration with equipment held in the lab. Realization of interfaces with DIH equipment. Provision of test benches (for actuation, perception, etc.)
Format (how?)	Conducting experiments with the assistance of DIH personnel.
Profile of DIH contact (who?)	Technical experts.
Involvement, effort from DIH	Support from experts in the submission of European or national proposals. Participation of several experts from the DIH for maturation.
Financial charge for customer	Yes: the price is based on a daily rate and depends on the personnel involved, equipment used at the DIH and the adaptation needed. Typically paid by national or European funding.
Funding, business plan	Funded by bilateral contracts, European or national innovation programmes, investors.
Certification, product demonstration & testing, comparison of products	
Category	Testing and validation.
Description of activity, value proposition (what?)	For testing and benchmarking, VTT and other DIH members can offer access to a set of dedicated infrastructures to enable technologies to be assessed in realistic conditions. DIH can organize access to infrastructure in many application domains.
Format (how?)	Conducting tests and experiments with the experts from VTT following evaluation protocols and applicable standards.
Profile of DIH contact (who?)	Technical experts.
Involvement, effort from DIH	Participation of several experts from the DIH, if needed.
Financial charge for customer	Yes.
Funding, business plan	Funded by bilateral contracts, European or national innovation programmes, investors.